Proceedings of the NEA Workshop on

SYSTEM PERFORMANCE ASSESSMENTS FOR RADIOACTIVE WASTE DISPOSAL

Compte rendu de la réunion de travail de l'AEN sur

L'ÉVALUATION DES PERFORMANCES DES SYSTÈMES D'ÉVACUATION DES DÉCHETS RADIOACTIFS

ORGA

ORGANIS

Pursuant to article 1 of the Convention signed in Paris on 14th December, 1960, and which came into force on 30th September, 1961, the Organisation for Economic Co-operation and Development (OECD) shall promote policies designed:

- to achieve the highest sustainable economic growth and employment and a rising standard of living in Member countries, while maintaining financial stability, and thus to contribute to the development of the world economy;
- to contribute to sound economic expansion in Member as well as non-member countries in the process of economic development; and
- to contribute to the expansion of world trade on a multilateral, non-discriminatory basis in accordance with international obligations.

The Signatories of the Convention on the OECD are Austria, Belgium, Canada, Denmark, France, the Federal Republic of Germany, Greece, Iceland, Ireland, Italy, Luxembourg, the Netherlands, Norway, Portugal, Spain, Sweden, Switzerland, Turkey, the United Kingdom and the United States. The following countries acceded subsequently to this Convention (the dates are those on which the instruments of accession were deposited): Japan (28th April, 1964), Finland (28th January, 1969), Australia (7th June, 1971) and New Zealand (29th May, 1973).

The Socialist Federal Republic of Yugoslavia takes part in certain work of the OECD (agreement of 28th October, 1961).

The OECD Nuclear Energy Agency (NEA) was established on 20th April 1972, replacing OECD's European Nuclear Energy Agency (ENEA) on the adhesion of Japan as a full Member.

NEA now groups all the European Member countries of OECD and Australia, Canada, Japan, and the United States. The Commission of the European Communities takes part in the work of the Agency.

The primary objectives of NEA are to promote co-operation between its Member governments on the safety and regulatory aspects of nuclear development, and on assessing the future role of nuclear energy as a contributor to economic progress.

This is achieved by:

- *encouraging harmonisation of governements' regulatory policies and practices in the nuclear field, with particular reference to the safety of nuclear installations, protection of man against ionising radiation and preservation of the environment, radioactive waste management, and nuclear third party liability and insurance;*
- *keeping under review the technical and economic characteristics of nuclear power growth and of the nuclear fuel cycle, and assessing demand and supply for the different phases of the nuclear fuel cycle and the potential future contribution of nuclear power to overall energy demand;*
- *developing exchanges of scientific and technical information on nuclear energy, particularly through participation in common services;*
- *setting up international research and development programmes and undertakings jointly organised and operated by OECD countries.*

In these and related tasks, NEA works in close collaboration with the International Atomic Energy Agency in Vienna, with which it has concluded a Co-operation Agreement, as well as with other international organisations in the nuclear field.

LEGAL NOTICE

The Organisation for Economic Co-operation and Development assumes no liability concerning information published in this Bulletin.

En vertu de l'article 1er de la Convention signée le 14 décembre 1960, à Paris, et entrée en vigueur le 30 septembre 1961, l'Organisation de Coopération et de Développement Économiques (OCDE) a pour objectif de promouvoir des politiques visant :

– à réaliser la plus forte expansion de l'économie et de l'emploi et une progression du niveau de vie dans les pays Membres, tout en maintenant la stabilité financière, et à contribuer ainsi au développement de l'économie mondiale ;
– à contribuer à une saine expansion économique dans les pays Membres, ainsi que non membres, en voie de développement économique ;
– à contribuer à l'expansion du commerce mondial sur une base multilatérale et non discriminatoire conformément aux obligations internationales.

Les signataires de la Convention relative à l'OCDE sont : la République Fédérale d'Allemagne, l'Autriche, la Belgique, le Canada, le Danemark, l'Espagne, les États-Unis, la France, la Grèce, l'Irlande, l'Islande, l'Italie, le Luxembourg, la Norvège, les Pays-Bas, le Portugal, le Royaume-Uni, la Suède, la Suisse et la Turquie. Les pays suivants ont adhéré ultérieurement à cette Convention (les dates sont celles du dépôt des instruments d'adhésion) : le Japon (28 avril 1964), la Finlande (28 janvier 1969), l'Australie (7 juin 1971) et la Nouvelle-Zélande (29 mai 1973).

La République socialiste fédérative de Yougoslavie prend part à certains travaux de l'OCDE (accord du 28 octobre 1961).

L'Agence de l'OCDE pour l'Énergie Nucléaire (AEN) a été créée le 20 avril 1972, en remplacement de l'Agence Européenne pour l'Énergie Nucléaire de l'OCDE (ENEA) lors de l'adhésion du Japon à titre de Membre de plein exercice.

L'AEN groupe désormais tous les pays Membres européens de l'OCDE ainsi que l'Australie, le Canada, les États-Unis et le Japon. La Commission des Communautés Européennes participe à ses travaux.

L'AEN a pour principaux objectifs de promouvoir, entre les gouvernements qui en sont Membres, la coopération dans le domaine de la sécurité et de la réglementation nucléaires, ainsi que l'évaluation de la contribution de l'énergie nucléaire au progrès économique.

Pour atteindre ces objectifs, l'AEN :

– *encourage l'harmonisation des politiques et pratiques réglementaires dans le domaine nucléaire, en ce qui concerne notamment la sûreté des installations nucléaires, la protection de l'homme contre les radiations ionisantes et la préservation de l'environnement, la gestion des déchets radioactifs, ainsi que la responsabilité civile et les assurances en matière nucléaire ;*
– *examine régulièrement les aspects économiques et techniques de la croissance de l'énergie nucléaire et du cycle du combustible nucléaire, et évalue la demande et les capacités disponibles pour les différentes phases du cycle du combustible nucléaire, ainsi que le rôle que l'énergie nucléaire jouera dans l'avenir pour satisfaire la demande énergétique totale ;*
– *développe les échanges d'informations scientifiques et techniques concernant l'énergie nucléaire, notamment par l'intermédiaire de services communs ;*
– *met sur pied des programmes internationaux de recherche et développement, ainsi que des activités organisées et gérées en commun par les pays de l'OCDE.*

Pour ces activités, ainsi que pour d'autres travaux connexes, l'AEN collabore étroitement avec l'Agence Internationale de l'Énergie Atomique de Vienne, avec laquelle elle a conclu un Accord de coopération, ainsi qu'avec d'autres organisations internationales opérant dans le domaine nucléaire.

FOREWORD

Currently one of the most important areas of radioactive waste management is research and development directed towards assessments of the performance of potential radioactive waste disposal systems. It is primarily for this reason that, under the guidance of its Radioactive Waste Management Committee (RWMC), the OECD Nuclear Energy Agency has devoted a considerable amount of effort towards the further development of systems performance assessment methodologies. The first NEA workshop on this topic, which was held in 1982, critically reviewed the status of performance assessment methodologies in both their generic and site-specific aspects in order to provide guidance for the further development of predictive mathematical models and also to facilitate comparison of different approaches to modelling and between safety assessments of alternative disposal options. The resultant informal report* identified the various approaches adopted in OECD Member countries, highlighted some of the problems foreseen in carrying out assessments and made recommendations on future needs.

In the interim since the first workshop, new challenges have arisen which stem from a concerted effort throughout the OECD area to develop and apply performance assessment methodologies, including some full scale assessments of potential disposal systems within several Member countries. A small group of consultants was assembled to consider these developments and, together with the NEA Secretariat, develop the most appropriate theme for a second NEA workshop. As a result, it was considered that due to the number of different approaches being followed, a major problem was how to rationalise all the various elements involved in carrying out performance assessments. This rationalisation was necessary in order to help generate confidence in predictions of the performance of waste disposal systems. It was therefore agreed that the workshop would examine the links or interrelationships between the major components of performance assessments, in particular the following :

- the link between performance assessments and regulatory requirements ;
- the links between mathematical models used in performance assessments ;
- the link between model development and field/laboratory observations.

These proceedings reproduce the papers presented at the Workshop together with a summary and conclusions prepared by the NEA Secretariat in conjunction with rapporteurs appointed for each session and the Chairman. The opinions expressed are the responsibility of the authors and it no way commit the Member countries of the OECD.

---

* This report is available on request from the OECD/NEA Secretariat, 38 boulevard Suchet, 75016 Paris, France.

AVANT-PROPOS

Les travaux de recherche et de mise au point consacrés à l'évaluation des performances des systèmes éventuels d'évacuation des déchets radioactifs occupent actuellement une place de premier plan dans le contexte de la gestion de ces déchets. C'est essentiellement pour cette raison que, sous l'égide de son Comité de la Gestion des Déchets Radioactifs (RWMC), l'Agence del'OCDE pour l'Energie Nucléaire s'est énergiquement attachée à perfectionner les méthodes d'évaluation des performances des systèmes. Les participants à la première réunion de travail de l'AEN sur ce thème, qui s'est tenue en 1982, ont procédé à un examen critique de l'état d'avancement des méthodes d'évaluation des performances, en analysant à la fois leurs aspects génériques et ceux propres à un site donné. L'objectif étant de fournir des orientations quant aux moyens d'améliorer les modèles mathématiques prévisionnels et aussi de faciliter les comparaisons entre les diverses démarches à l'égard de la modélisation et entre les évaluations de la sûreté d'autres systèmes d'évacuation. Le rapport officieux publié à l'issue de cette réunion* recense les diverses démarches adoptées dans les pays Membres de l'OCDE, met en relief quelques-uns des problèmes anticipés dans l'exécution des évaluations et contient des recommandations concernant les besoins futurs.

Depuis cette première réunion de travail, de nouveaux problèmes ont surgi à la suite de l'effort concerté qui a été déployé dans l'ensemble de la zone de l'OCDE en vue de mettre au point et d'appliquer des méthodes d'évaluation des performances, y compris quelques évaluations en vraie grandeur de systèmes d'évacuation potentiels dans plusieurs pays Membres. Un petit groupe de consultants a été convoqué pour examiner ces initiatives et, de concert avec le Secrétariat de l'AEN, définir le thème le plus approprié pour une deuxième réunion de travail de l'AEN. La conclusion qui s'est dégagée de ces travaux est que, vu le nombre de démarches adoptées en l'occurrence, la rationalisation de l'ensemble des divers éléments qui entrent en ligne de compte dans la réalisation des évaluations des performances constitue un problème de premier plan. Cette rationalisation est nécessaire pour contribuer à susciter de la confiance dans les prévisions relatives aux performances des systèmes d'évacuation des déchets. Il a donc été convenu que les participants à la réunion de travail examineraient les relations ou l'interdépendance entre les principaux éléments qui interviennent dans les évaluations des performances, et notamment les suivants :

- la relation entre les évaluations des performances et les prescriptions réglementaires ;

- les relations entre les modèles mathématiques utilisés dans les évaluations des performances ;

---

* Ce rapport est disponible sur demande auprès du Secrétariat de l'Agence de l'OCDE pour l'Energie Nucléaire, 38 boulevard Suchet, 75016 Paris, France.

- la relation entre la mise au point des modèles et les observations sur le terrain et en laboratoire.

Le présent compte rendu contient le texte des communications présentées à la réunion de travail, ainsi qu'un résumé et des conclusions établis par le Secrétariat de l'AEN avec le concours du Président et des rapporteurs désignés pour chaque séance. Les opinions exprimées représentent le point de vue de leurs auteurs et n'engagent en aucune manière les pays Membres de l'OCDE.

TABLE OF CONTENTS
TABLE DES MATIERES

Session III - Séance III

LINKAGES BETWEEN MODELS IN SYSTEM
PERFORMANCE ASSESSMENTS

Session IV - Séance IV

THE LINK BETWEEN MODEL DEVELOPMENT AND
FIELD/LABORATORY OBSERVATIONS

EXECUTIVE SUMMARY AND RECOMMENDATIONS

INTRODUCTION

At the fifteenth meeting of the NEA Radioactive Waste Management Committee in February 1985, a review of the Committee's future activities was carried out. As a consequence, it was agreed that a major part of its work should be focussed on ways to help further the development of system performance assessment methodologies for use in carrying out safety assessments of potential radioactive waste disposal facilities. In order to concentrate on the most appropriate topics, it was also agreed that a second workshop should be held in October 1985 on the development and application of such systems performance assessment techniques. It was the intention that the workshop would build on the findings of the first NEA workshop on this topic, which was held in November 1982, and experience gained in the interim period and would address the most pressing issues in this field.

The first NEA Workshop on "System Performance Assessments on the Safety of Radioactive Waste Disposal" identified and reviewed most of the elements involved in carrying out comprehensive assessments of the radiological impact, or "performance", of a range of disposal options for radioactive waste. The resultant report* identified the various approaches adopted in Member countries and highlighted some of the problems foreseen and where future work should be focussed. Since that workshop, a great deal of effort has been devoted to the further development and application of performance assessment methodologies and the provision of reliable data. This led to new challenges arising which provided the main stimuli for a second workshop to be held in October 1985, the focus of which would be the current problem areas and ways that these can be solved, especially at an international level. A small consultant group**, which met with the Secretariat to develop the theme and detailed programme, considered that the increasing number of methodologies, models, criteria, etc., had led to some concern among OECD Member countries on how performance assessments should be carried out. It was suggested that there exists one particular problem that should be resolved with the aid of discussions at an international level, i.e. how to rationalise all the various elements that combine together in carrying out performance assessments. This rationalisation was thought necessary in order to help promote confidence in the reliability of performance assessments and in the ability to provide the correct type of information on which the efficiency of disposal can be judged. It was further considered that fundamental to this rationalisation is an awareness of the major relationships

---

* Available from the OECD Nuclear Energy Agency.

** Participants in the NEA consultant group on System Performance Assessments: Dr. P.D. Johnston (DOE, UK), Mr. R.B. Lyon (AECL, Canada), and Dr. T. Papp (SKB, Sweden).

or linkages between the major elements involved in carrying out and utilising performance assessments. Figure 1 gives a schematic representation of the main linkages in system performance assessments. The group recommended that three specific interrelationships should be examined, i.e.

Figure 1

LINKAGES IN SYSTEM PERFORMANCE ASSESSMENTS

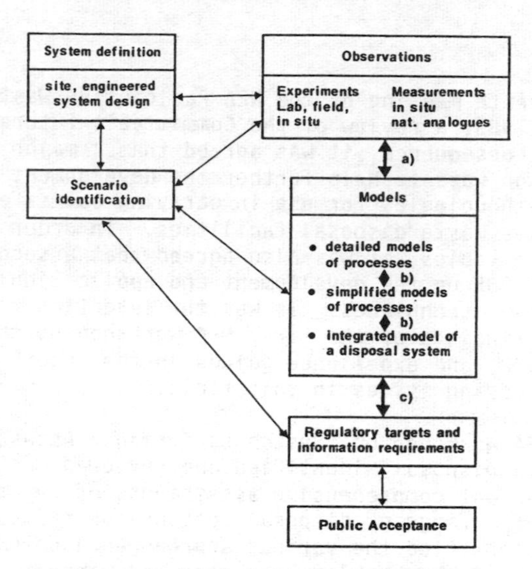

a) Link between the development of models and observations (validation)
b) Link between detailed models and simple models and Link between separate models and an integrated system model
c) Link between the output of performance assessments and regulatory requirements.

a) The link between mathematical models and field and laboratory observations.

b) The link between the various mathematical models used in performance assessments, both i) between detailed (research) models and simplified models and ii) between models in an integrated systems model.

c) The link between the output from performance assessments and regulatory requirements.

Such linkages have often been identified as weak points in current assessments and are of common interest across the spectrum of disposal options. The objective of the workshop was therefore set as:

- to examine the main linkages between particular elements of system performance assessments for radioactive waste disposal, to identify areas where improvements can be made and suggest ways of carrying these out.

In trying to meet this objective, particular emphasis was placed on the possible future initiatives that could be taken by the Nuclear Energy Agency. The following text is a summary of the most important points arising from consideration of each linkage during the workshop. It is based on reviews by rapporteurs appointed for each session*.

THE LINK BETWEEN PERFORMANCE OBJECTIVES AND PERFORMANCE ASSESSMENTS

Arising from the presentations and discussion of the linkage between performance assessments and regulatory requirements, it was generally agreed that performance objectives can be formulated in a number of different ways and yet still ensure that radioactive wastes are disposed of safely. General objectives relating to radiation dose, health risk (to individuals or populations) or to environmental contamination have been proposed and are under discussion. There appears to be a trend towards individual risk limitation in several European countries and in Canada, which are similar to the recommendations of an NEA Expert Group on Long-Term Radiological Protection Objectives for Radioactive Waste Disposal and more recently to principles published by the ICRP. Other European countries may prefer a system of individual dose limitation for reasonable scenarios of radiation exposure. In the United States, the regulatory bodies (EPA and NRC) have developed limits on radionuclide releases to the environment and also criteria for hydrogeological conditions and the performance of the waste package such that population health risk objectives will be met. Alternatively, the Swedish authorities may give greater emphasis to comparison with natural fluxes of radioactivity in the environment in judging radiological safety in the very long term.

The way the link is made to performance assessments, and the way assessments are carried out, will depend on the type of objective adopted in each country and on national regulatory mechanisms. In waste repository siting and design, general radiological objectives will need to be translated into geological, hydrogeological and engineering specifications. In many countries, this will be left to the developers of repositories: specifications may be used to focus research and design programmes, they may be site specific and may evolve as site investigations progress, or they may serve simply to define an envelope of conditions within which developers feel confident that regulatory authorities can be convinced that general objectives are satisfied.

Despite these differences, performance assessment procedures have many similarities. There is a universal awareness of the need for uncertainty analysis, and an increasing use of probabilistic assessment procedures: there is everywhere a recognition of the need for transparency in procedures and a need to validate models wherever practicable. A number of common problems emerged in the discussions and are listed below.

---

* Rapporteurs for each session were: Session II, Dr. P.D. Johnston (DOE, UK); Session III, Dr. A. Saltelli (JRC, CEC); Session IV, Dr. R.M. Cranwell (Sandia Laboratories, USA).

1. Identification and Screening of Scenarios

While there can be no absolute guarantee that all important disposal scenarios have been assessed, such a wide range of possible scenarios have been identified that the probability of missing an important scenario is extremely low. The scenarios chosen will be a function of the site and repository design. An initial screening could eliminate those with such low probability that a risk limit, or a "reasonable scenario" dose limit, would be respected for any feasible consequence. A further screening on the basis of risk or consequence can then allow effort to be concentrated on the few sequences of events or processes of greatest potential importance. General screening procedures might be developed, although the list of scenarios retained for detailed study will be site specific.

2. Assignment of Probabilities to Scenarios and of Probability Distributions to Modelling Parameters

This is currently an area of major importance. The difficulties of assigning probabilities and probability distributions are balanced by the ability to incorporate uncertainty, the latter being essential for projections into the future. Expert judgement will be necessary when historical data are unavailable or where probability distributions cannot be derived from experiment or detailed modelling and sensitivity analysis. In such cases it is often difficult to demonstrate "reasonableness" in the choice of data and probabilities and a sharing of experience at an international level could be very valuable in this area.

3. Establishing Confidence that Disposal Facilities will be Safe

This is as much a problem for regulatory bodies in setting limits as it is for developers of facilities. Transparent procedures, a recognition of uncertainties, independent assessments and peer review will all be necessary for establishing confidence. It may also be valuable to present information in different ways to different audiences, and it may therefore be advisable not to focus assessments exclusively on demonstrating compliance with technical regulatory objectives.

A risk limit would be unaffected by changes in the understanding of the link between radiation dose and health risk. Risk limitation may therefore be a robust regulatory practice which could avoid the need for periodic changes in legislation, with consequent damage to public confidence.

4. Co-operation Between Assessment Groups and Regulatory Authorities

It is essential that experts involved in performance assessments and the staff of regulatory bodies work together to ensure that regulations are formulated such that they can be shown to be respected, and so that assessments generate appropriate information. Joint projects like the CEC PAGIS project can be a valuable common forum for discussion.

THE LINK BETWEEN DETAILED PROCESS MODELS AND SIMPLIFIED MODELS

Simplification is particularly important when linking process models together in assessments of a total disposal system, i.e. including the waste form, vault, geosphere and biosphere. This is because combining detailed process models is very complicated and the resultant model would be cumbersome, difficult to change and somewhat opaque to review. Further, their use would stretch computer resources and the quality of the resultant performance assessments would probably not be significantly enhanced by the increased detail.

A great number of processes have been identified that might have an impact on the safety of a waste disposal repository. Some of these processes are well understood and validated models exist, such as for heat transfer; others are much less well known and are only possible to model in very simplistic ways, such as earthquake and glaciation phenomena. Ideally, research into the less predictable phenomena should yield greater understanding and sooner or later a valid detailed research model of the process. If this detailed model is too complex for use in global assessments, then a simplified model would be derived which would:

a) use less computer time, and
b) retain the essence of information relevant to the assessment.

The session on the link between detailed and simplified models demonstrated that systematic methods can be used to simplify detailed models. Areas requiring further attention are listed below.

1. The Low Quality of Some Research Models

This is particularly the case for the poorly understood phenomena, such as flow within fractures. In these cases either research effort has to be orientated towards increasing understanding, or a disposal system should be designed to limit the influence of the phenomenum on performance. An example of the latter could be in cases where earthquakes are possible, a site could be selected such that it was surrounded by zones of structural weakness which would absorb the shock of an earthquake and leave the central block (with the repository) unaffected. An advantage of this site-specific design-away-the-problem approach, is that uncertainties in the performance of a waste disposal system are minimised and hence confidence is increased.

2. The Range of Validity of Simplified Models

An effect of using simplified models with a parameter span of limited validity is that it will be necessary to show that the interesting or crucial parts of the calculated consequence spectrum for the repository, really are within the span of the parameters chosen. In such cases, it may be advisable to make specific deterministic runs using the research models on those areas of interest.

3. Biosphere Models Require Special Consideration

The development of a long-term biosphere model is especially difficult because of the relatively rapid changes that occur in the biosphere. There was some debate on how this should best be undertaken, for example, whether this simplification should be done by weighting a variety of possible biosphere situations with their probability of occurrence, or by selecting a standard environment. Another approach would be to exclude the biosphere from the analysis and base the acceptability on releases from groundwater to the biosphere. This area clearly needs to be considered further at an international level.

THE LINK BETWEEN INDIVIDUAL PROCESS MODELS AND INTEGRATED SYSTEMS MODELS

Examination of the link between individual process models concentrated on probabilistic and deterministic assessment methodologies. It was clear from the presentations and discussions that the interrelationships are well understood for both types of code in that the current capabilities and problem areas are well defined. Areas requiring further attention are listed below.

1. Boundary Conditions

It was recognised that separating processes into individual submodels is achieved at the expense of approximations in the boundary regions between linked submodels. The general recommendation which can be made in this context is that the error should be forced in a pessimistic direction, i.e. the approximation must be conservative.

In Probabilistic Systems Assessment Codes (PSACs), the assumption of zero downstream interfacial concentration can be made whenever transport is faster in the downstream module than it is in the upstream one. According to such a condition, flux continuity is maintained across the two media but concentration continuity is not. This condition is conservative and has been extensively employed e.g. (NUTRAN, SYVAC 1). A refinement of this condition is proposed in SYVAC 2 (AECL, Canada), where flux out of the buffer is equated to a function of the relative magnitude of the transport in the geosphere and in the buffer.

In Deterministic Assessment Codes (DACs), a more accurate method of linking transport models is proposed in MELODY (France) where concentration continuity across the boundary is achieved by iterative interpolation at the interface using the Newton method. This technique is both accurate and effective but is likely to be too time consuming for use in PSAC.

2. Testing of Codes and Quality Assurance

Testing of the integrated system must be extensively carried out in order to ensure that the proper links have been established between the various submodels. Preliminary testing of the individual submodel is recommended following established procedures such as:

i) Each module has to be tested as a "stand alone" before inserting it in the code.

ii) The test must involve the same wide range of input parameters which the module is expected to receive in the code.

iii) If the module is a mass transfer one (e.g. buffer, geosphere, etc.), it must be extensively tested for mass balance.

Standard techniques that can be recommended for the testing of the integrated system are Quality Assurance (QA) and intercomparisons. QA is a very promising technique provided it is adapted to the rapid evolution of the assessment codes submodels. Intercomparisons at the level of the integrated systems are presently underway and the future results will tell us how far these exercises can assist in model validation and confidence building.

Apart from these validation procedures there are a number of tests that can be easily applied to the integrated system. The following is a list of simple operations that can be recommended:

i) Dimensionality check: starting from the moles of a given element per unit of waste form, check that the proper units for the output consequence are obtained.

ii) Testing of the system for limiting values of the input parameters, for instance: set the inventory of a parent nuclide to zero and check that the element does not show up in the output. Set the decay constant of an element to zero and check that its daughters do not appear in the output. Similarly, solubilities, consumption rates, transfer factors, can be set to zero to make sure that the system is working properly.

iii) Back of the envelope calculation: the dose resulting from one or more significant runs can be checked -- to an order of magnitude - by roughly estimating transit times through the different barriers and the attenuation of the pulse. In doing so, it can be useful to neglect dispersivity - where possible and to assume the steady-state for the transfer between the various compartments.

3. <u>The Role of Correlation</u>

This issue was touched upon only in relation to PSAC. Correlation techniques are in fact especially needed to mitigate the effect of the random choice of correlated input parameters.

A review of the existing methodologies resulted in the following points being made:

i) The basic variable approach foreseen for SYVAC-3 appears elegant and correct, although it might be very demanding in terms of modelling the relationships between the dependent parameters and the basic set.

ii) The method presently used in SYVAC-2 is only applicable to normal and log-normal distributions.

iii) The Iman method implemented in LISA is both flexible and general (distribution free).

iv) Alternatively, the sampling can be forced in corresponding quartiles for the variables to be correlated (CRNL code ECHOS). This approach is simple and effective. Although it does not yield a pre-specified correlation, it can eliminate unrealistic parameter combinations.

4. Optimisation of Calculations

The use of detailed models in the deterministic approach presents, a priori, the danger of developing an excessively large computer tool which may be time-costly, uneasy to manage and ill-fitted to model evolution following the introduction of new models. That is why the optimisation of calculations in both its numerical as well as its informatic aspects is to be considered very soon in France.

Apart from the choice of the level of model complexity, previously discussed, this optimisation in particular needs to take account of:

i) The separation of the different tasks involved, i.e., data management, chaining of calculations, performing calculations, results management, the modularity of the global structure and the associated documentation of rules for computational modules.

ii) The introduction of safeguards-restart procedures.

iii) The use of different time steps according to the different models.

THE LINK BETWEEN MODEL DEVELOPMENT AND FIELD/LABORATORY OBSERVATIONS

It was generally agreed that in order to achieve the predictive capability required to carry out comprehensive performance assessments, it is essential to develop a thorough understanding of the processes involved, fully characterise the system being modelled and possess a complete data base for use in making predictions. With this in mind, the key issues raised in terms of understanding the linkage between model development and field/laboratory observations were:

i) How to make reliable predictions of future behaviour?

ii) How to take account of uncertainty in predictions? and

iii) When has sufficient site characterisation been achieved?

The linkage between model development and field/laboratory observations was felt to represent an interactive process of site and system characterisation. Initially, the available system characterisation data are used to develop preliminary models for use in setting initial performance allocation goals, designing and directing further experiments to improve the site characterisation data base and alter, if necessary, the conceptual models. Ultimately, the goal becomes one of developing an observational data base that supports the validity of the detailed subsystem models for the range of conditions that are important to performance assessment needs, and to determining when the performance goals are achieved.

1. Model Validation

The problem here arises from the fact that "full validation", in the context of complete confirmation of future behaviour, can never be achieved. Laboratory and field experiments, properly conducted and carefully designed, are crucial for model validation. Several problem areas associated with laboratory and field experiments were identified. These were: 1) sampling procedures can alter the properties of samples; 2) time and spatial scales for experiments are short compared to time and spatial scales involved with repository performance assessments; 3) difficulty in simulating conditions at repository depths; 4) insufficient data from test sites; and 5) uncertainties concerning groundwater flow.

Information from natural analogues is also of major importance for model validation, especially with respect to long time scales. Problems with natural analogues, however, arise from the lack of good analogues in the time range of 1000 to 100 000 years and in defining initial boundary conditions. Also, almost all useful information which can be obtained from natural analogues relates to chemical processes and is of little use in groundwater flow modelling.

Possible Solutions

It was considered valuable to develop an international consensus on a strategy for validation with the objective being to reach some agreement on the range of applicability of different modelling approaches and reasonable assurance that the models provide a good representation of the processes occurring. The degree of validation would be different for different models depending on their role in performance assessment. The need for more carefully designed experiments for the purpose of model validation was also suggested. It was felt that to achieve this, there was a need for close collaboration between field and laboratory experimentalists, geologists and modellers. Further investigation into the use of natural analogues was recommended. Finally, international benchmarking programs such as INTRACOIN and HYDROCOIN were felt to be extremely useful for addressing the problems of model validation. The benchmarking program INTRAVAL and that proposed within the NEA User Group for Probabilistic Systems Assessment Codes will also be very useful in the model validation effort.

2. Data

Concerns about performance assessment data can best be classified as 1) availability of data; 2) acquisition of data; and 3) use of data. Problems with the availability of data arise from the degree to which data will be available for model development (for example, fracture data for use in a dual-porosity model) and the bias injected by overlooking the original purpose of previous data collection efforts. For example, oil well exploration data provide a convenient source of existing data on deep systems but the drill stem testing techniques commonly used to measure hydraulic properties have the potential for routinely excluding any of the higher permeability measurements because of the limitations of the technique. Identification of bias is important to both the use of existing data in models and the design of new data collection efforts.

Problems with data acquisition can arise from the sample size, frequency (spatial variation), the tools and instruments used to collect data, and interpretation and extrapolation of data. For example, observations and measurements of parameters are made at "points" within the system. However, characterisation of the variability of these parameters in space and time is typically required to model and make performance assessment predictions. Thus, the "point" information needs to be extrapolated over the spatial and time domains. A more complicated situation arises for parameters that cannot be measured directly (e.g. permeability and dispersivity), but must be determined indirectly through inverse modelling techniques.

Problems associated with the use of data can arise from the misuse of previously collected data, (as discussed above), use of so-called "lumped" parameters (e.g. distribution coefficients), and use of homogeneous data in a heterogeneous system.

Possible Solutions

Careful use of data, improved measurement techniques, close collaboration between experimentalists and modellers, and well-defined data acquisition programs were all suggested as possible solutions to the performance assessment data problem. Issues that need to be considered and addressed when field data are used and when planning and designing data acquisition programs were suggested. They include:

- How should small-sample data be averaged to obtain equivalent large-sample estimates for our performance assessment models? Is it necessary, and is it appropriate?
- What effect does variability in sample size have on our ability to obtain estimates for the spatial distribution of the data set, and what effect will this have on our estimates of spatial correlation lengths?
- How important is it that many of the parameter interpretation theories were developed for a homogeneous world while the real world is heterogeneous?
- For inversely determined parameters in a heterogeneous world, what is the appropriate relationship between:
  - the perturbation stimulus;
  - kind, number, locations, and sampling size of response observations;
  - model used for test interpretation;
  - the sampling size of the test; and
  - the band width of spatial frequencies the test can detect.

Other issues of importance when using data and when planning data acquisition programs were:

- Purpose of the assessment and stage of the assessment program.

- Conceptual model or models for the system.

- Performance assessment approach (e.g. detailed or bounding) and the theory associated with this approach.

- Kinds of tools or instruments used to gather the data or make observations.

- Methods used to interpret and extrapolate these measurements or data.

3. Uncertainty

Several sources of uncertainty in performance assessment were identified. These included 1) data; 2) models; 3) human error; 4) future events; 5) time and spatial scale effects; and 6) understanding basic physical and chemical processes. It was felt that a major effort in developing confidence in our performance assessment predictions would be in reducing, quantifying, or bounding the uncertainties associated with all important components involved in making performance assessment predictions.

Possible Solutions

Uncertainty analysis should be an integral part of any performance assessment methodology, regardless of the performance objective required by the regulatory agency. Several techniques for performing uncertainty analysis currently exist. Some of the more commonly used are 1) the classical "Monte Carlo" simulations; 2) differential analysis techniques; and 3) experimental design methods. Other more recent approaches are 1) the development of stochastic models; 2) geostatistical methods such as kriging; and 3) so-called statistical inverse methods. Additional work in this area needs to be encouraged, such as that proposed in the Level 3 HYDROCOIN program and international workshops on uncertainty analysis. The recent activities of the NEA in the formation of a consultant group on uncertainty analysis is a step in the right direction.

RECOMMENDATIONS

The final plenary session of the workshop was devoted to presentations by the rapporteurs for each session, followed by open discussion of future needs in the area of system performance assessments. Particular emphasis was placed on possible initiatives to be addressed by the NEA. Arising from this, a number of suggestions were made and, following subsequent deliberations within the NEA Secretariat, a number of broad recommendations were formulated.

- The NEA should be encouraged to host a workshop on uncertainties in performance assessments. This reflected discussions at the Workshop, but also endorsed the recent activities of the RWMC in this area, where a consultant group has produced a preliminary report on Handling Uncertainties in System Performance Assessments.

- The NEA should be encouraged to host a workshop or establish an expert group to develop an agreed international methodology for scenario identification. Particular emphasis should be given to a) the screening of scenarios; b) the assignment of probabilities to scenarios; and c) assignment of probability distributions to modelling parameters. A specific topic to be addressed would be the identification of scenarios covering biosphere evolution and the development of a methodology for the treatment of this evolution in performance assessments.

- The NEA should consider the continuation of the topical workshops on performance assessments. A possible topic for the next workshop is "Strategies for Confidence Building", which would cover validation, peer review, public consultation and communication, quality assurance, etc.

- The NEA should be encouraged to examine ways to respond to requests for international peer review of performance assessment activities, ranging from overall assessments (such as the review of KBS-3) to specific components (such as that provided by the Users Group for Probabilistic Systems Assessment Codes).

- The NEA can play a valuable role in the area of code development by providing an exchange facility via a model (code) library established at the NEA Data Bank.

- The intercomparison of codes was endorsed by the group as being necessary for their verification. The NEA should be encouraged to play a role in the co-ordination of such activities. A suggested new activity was the establishment of a group to compare near-field (engineered barrier) codes.

- The NEA's active role in the development of long term radiological criteria was endorsed.

These and other recommendations were considered by the Radioactive Waste Management Committee at its sixteenth meeting in December 1985 where a number of initiatives were agreed: a workshop on uncertainty analysis; a workshop or expert group to develop an agreed international methodology for scenario identification; and to establish an NEA Performance Assessment Advisory Group. The latter was agreed upon, following a recommendation from the NEA Secretariat, to provide advice on performance assessment-related matters and help co-ordinate the many NEA activities in this field.

SESSION I

THE STATUS AND DEVELOPMENT NEEDS OF SYSTEM PERFORMANCE ASSESSMENTS

THE IMPORTANCE OF LINKAGES IN POST-CLOSURE SYSTEM PERFORMANCE ASSESSMENTS

IMPORTANCE DES LIAISONS CONCERNANT L'EVALUATION DES PERFORMANCES DES SYSTEMES D'EVACUATION DES DECHETS RADIOACTIFS APRES FERMETURE DES DEPOTS

S. G. Carlyle
Nuclear Energy Agency
Organisation for Economic Co-operation and Development
Paris, France

ABSTRACT

This paper briefly reviews the findings of an NEA Consultant meeting held on 29-30 April 1985 which met to consider the programme for a workshop on system performance assessments. It describes the rationale behind the choice of linkages between different elements of performance assessment as the focus of the workshop, and the topics chosen for review.

RESUME

Cette communication résume brièvement les résultats de la réunion de consultants de l'AEN qui s'est tenue les 29-30 avril 1985 en vue d'examiner le programme d'une réunion de travail sur l'évaluation des performances des systèmes. On y trouvera une description du thème de la réunion sur la justification du choix des liaisons entre les différents éléments de l'évaluation des performances, ainsi que des autres sujets examinés.

Introduction

In order to fully understand the aims of this workshop on system performance assessments, it is first of all necessary to be aware of the circumstances which led to its conception. It was organised by the NEA following the agreement of the Radioactive Waste Management Committee (RWMC) at its meeting in February 1985. The RWMC has three main objectives:

1. to promote studies and improve the data base available to back up national radioactive waste management and demonstration programmes;

2. to contribute to the effectiveness of R & D in support of national radioactive waste management programmes and policies; and

3. to assist member countries in improving the level of understanding of radioactive waste management issues and options, particularly in the field of waste disposal.

It can be seen that this workshop is part of an ongoing programme of work designed to meet these aims. In fulfilling these objectives several important studies have recently been carried out by the NEA which led to the publication in 1984 of three notable reports, i.e.:

- Long Term Radiation Protection Objectives for Radioactive Waste Disposal;

- Geological Disposal of Radioactive Waste: An Overview of the Current Status of Understanding and Developments (a joint report with the CEC);

- Seabed Disposal of High Level Radioactive Waste: A Status Report on the NEA Co-ordinated Research Programme.

The conclusions and recommendations from these reports, together with those from other NEA activities, for example, on sea dumping and uranium mill tailings, were used as the basis for a "Technical Appraisal of the Current Situation in the Field of Radioactive Waste Management - A Collective Opinion by the Radioactive Waste Management Committee" which was published in January 1985. As the title suggests, the report drew conclusions on the status of understanding on current issues in the field of radioactive waste management. The fundamental conclusion of this "collective opinion" was that detailed short and long term safety assessments can now be made which give confidence that radiation protection objectives can be met with currently available technology for most waste types. It was also emphasized that R & D will have to continue to fill remaining gaps for particular disposal options, to collect site-specific data and refine safety studies.

The "collective opinion" report has proved to be a major milestone in the activities of the NEA in the area of radioactive waste management. It prompted a review of the future work of the RWMC with the result that the Committee agreed that three areas of activity would be covered in its programme of work, i.e. (i) waste management policies and strategies;

(ii) the assessment of the safety of radioactive waste management options, in particular in the long term; and (iii) various topical activities. The first activity concerns discussions at Committee level of policy orientated activities such as international co-operation on disposal and reviews of national programmes. The third activity covers several individual self-contained projects such as studies on the decommissioning and decontamination of nuclear facilities and on uranium mill tailings management. The second activity was seen as the most important and the area where most effort was needed. This confirmed an established trend in the work of the RWMC where several substantial activities were under way, for example the NEA Probabilistic Systems Assessment Codes (PSAC) User Group, the HYDROCOIN exercise, the International Sorption Information Retrieval System (ISIRS) and the Thermochemical Data Base.

In order to focus work on the most appropriate topics in this field it was also agreed that a workshop should be held in October 1985 on the development and use of system performance assessment techniques in judging the safety of radioactive waste disposal. The workshop would build on the findings of the first NEA workshop on this subject in November 1982 and experience gained in the interim period, and would consider the most pressing issues currently prevailing in this field. It was further agreed that a small group of consultants* would work with the NEA Secretariat to highlight the key topics to be addressed at the workshop and prepare a detailed programme.

The Objectives of the Workshop

The first NEA Workshop on System Performance Assessments identified and reviewed most of the elements used to make comprehensive radiological assessments of the impact of disposal. The resultant report identified the various approaches adopted in Member countries and highlighted some of the problems foreseen in carrying out assessments and future needs. In the interim since the workshop a great deal of effort has been devoted in OECD countries to the further development of assessment methodologies and the provision of reliable data. New challenges have arisen which provide the main stimuli for a second workshop in 1985 which would build on the findings of the 1982 workshop and more recent experience.

It was considered that the focus of the workshop should clearly be on current problems in carrying out comprehensive assessments of the performance of waste disposal systems. The increasing number of methodologies, models and criteria has led to some confusion on how such assessments should be carried out. It was suggested that there exists one particular problem which requires resolution, i.e. how to <u>rationalise</u> all the various elements that combine together in carrying out performance assessments. This rationalisation is necessary in order to generate <u>confidence</u> that performance assessments are reliable, realistic and can provide the correct type of information on which

---

* Participants in the NEA consultant group on system performance assessment were: Dr. P. D. Johnstone (DOE, UK), Mr. R. B. Lyon (AECL, Canada) and Dr. T. Papp (SKB, Sweden).

Figure 1

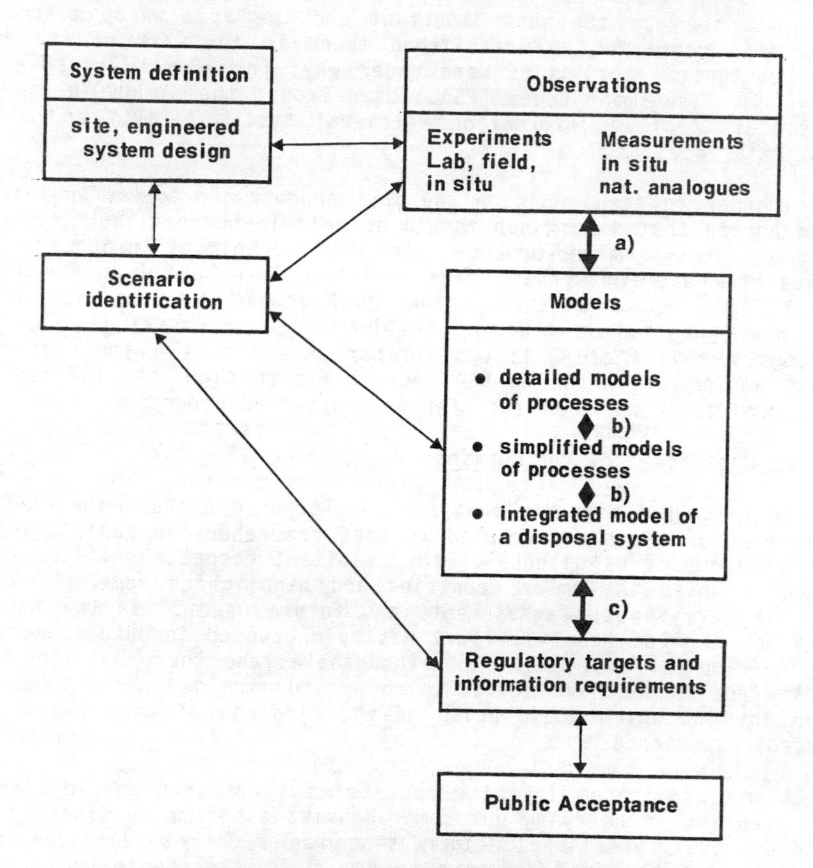

a) Link between the development of models and observations (validation)
b) Link between detailed models and simple models and Link between separate models and an integrated system model
c) Link between the output of performance assessments and regulatory requirements.

the efficacy of disposal can be judged. Fundamental to this rationalisation to breed confidence is an awareness of the various <u>links</u> between each component of the system. Different types of linkage exist such as the ones between those utilising and those carrying out performance assessments, through the link between those acquiring data and those utilising the data in models, to the way individual component models may be coupled together. It was considered by the consultant group that such linkages are a weak point in current assessments and are of common interest in assessments for a wide range of wastes and disposal options. In addition, by examining these linkages it should be possible to identify areas requiring greater attention and further work. Therefore it was decided that the objective of the workshop should be to examine the main links between particular elements of system performance assessments for radioactive waste disposal facilities, to identify areas where improvements can be made and suggest ways of carrying these out.

In meeting the above objective the workshop would cover all types of radioactive wastes, all disposal options and emphasis will again be placed on long term aspects of performance assessments. It would also provide an opportunity for a free exchange of views between experts representing research, design, engineering, regulation and licensing organisations from OECD countries.

<u>Topics and Scope of the Workshop</u>

After considerable discussion in the consultant group of the many links between specific elements of a performance assessment system, three principal linkages were identified as meriting most attention, i.e. (i) the link between the output from performance assessments and the needs of regulators; (ii) the link between the various predictive mathematical models used in performance assessments; and (iii) the link between model development and field/laboratory observations. There are several reasons why these linkages were identified as being of particular importance. The first linkage was thought of crucial importance as it is essential that performance assessors fully understand the needs of regulators and that regulators understand the capabilities and limitations of performance assessments involving the use of predictive mathematical models. The second linkage was considered of concern due to the various types of computer-based models used in performance assessments, each with specific characteristics that require detailed understanding when used in conjunction with other models. The third linkage was considered important as it covers the interpretation of field observations of heterogeneous systems by using a collection of mathematical formulae in computer-based models, one particular problem being the validation of these models by subsequent field observations. Other links were discounted by the group as being less problematic or because they are currently being considered in other fora. Figure 1 illustrates the main linkages and on this basis the following topics were included in the programme of the workshop:

Session I THE CURRENT STATUS AND DEVELOPMENT NEEDS IN SYSTEM PERFORMANCE ASSESSMENTS

- The Importance of Linkages in Post-Closure System Performance Assessments: Review of the Conclusions of the Consultants Group

- The Present Status and Current Challenges in System Performance Assessments; covering recent assessments for a range of waste types and disposal concepts

Session II THE LINKAGE BETWEEN PERFORMANCE ASSESSMENTS AND REGULATORY REQUIREMENTS

- The Link Between General Performance Objectives and Performance Assessments; covering presentation of information in PAGIS and European regulatory requirements; AECB requirements and output from SYVAC; UK DOE requirements for ILW, etc.

- A Prescriptive Approach; the Link Between the Performance Assessments of the US DOE Projects and Regulatory Reauirements for NRC and EPA

Session III LINKAGES BETWEEN MODELS IN SYSTEM PERFORMANCE ASSESSMENTS

- The Link Between Detailed Process Models and Simplified Models; covering the need for simplification, how can simplified models be generated, the role of sensitivity analysis in linkages and verification of simple models against detailed models.

- The Link Between Individual Process Models and Integrated Models of a System, covering correlations of processes/parameters; boundary conditions; examples of integration in probabilistic assessment codes in the USA, Canada, UK and the CEC; and the integration of detailed models in deterministic assessment codes in France and the FRG.

Session IV THE LINKS BETWEEN MODEL DEVELOPMENT AND FIELD/LABORATORY OBSERVATIONS

- Linkage Between Laboratory/Field Observations and Models; covering the availability of data, the use of field data in models, planning data acquisition programmes and adapting models to data availability

- Prospects for Model Validation Against Field/Laboratory Observations including a review of methods used and lessons learnt in HYDROCOIN and INTRACOIN and the use of natural analogues.

The formal sessions would be followed by a plenary session where rapporteurs would make brief presentations on the main points arising from the formal presentations and discussions with a view to generating further general discussion on future needs. Particular reference would be made to possible initiatives that could be taken by the NEA.

PRESENT STATUS AND CURRENT CHALLENGES IN SYSTEM PERFORMANCE ASSESSMENTS

C. McCombie
National Cooperative for the Storage of Radioactive Waste (NAGRA)
Baden, Switzerland

ABSTRACT

This paper reviews the present status of system performance assessment. Factors influencing the methodology applied in an assessment include the aims, the safety criteria and the timescales. These are discussed using examples of published assessments. The current consensus on the rôles, capabilities and problems of performance assessment is summarized. Some special remarks are included on the status of probibilistic methodology before concluding with an abbreviated list of current challenges.

L'EVALUATION DES PERFORMANCES DES SYSTEMES : ETAT D'AVANCEMENT DES TRAVAUX ET TACHES A ACCOMPLIR

RESUME

Ce document fait le point de l'état d'avancement des travaux relatifs à l'évaluation des performances des systèmes. Les facteurs qui influencent la méthodologie mise en oeuvre dans une évaluation comprennent notamment les objectifs visés, les critères de sûreté et les échelles de temps. Les auteurs analysent ces facteurs en s'appuyant sur des exemples d'évaluations publiées. Ils font la synthèse de l'opinion qui prévaut généralement quant au rôle de l'évaluation des performances, à ses possibilités et à ses problèmes. La communication contient quelques remarques spécifiques sur l'état d'avancement de la méthode probabiliste et se termine par une liste abrégée des tâches à accomplir.

1 THE ROLES OF PERFORMANCE ASSESSMENT (P.A.)

In developing concepts and technology for safe disposal of radioactive wastes one is confronted with the problem of predicting the behaviour of relatively complex technical systems over long time periods. This is a novel task. Very much more complex systems (e.g. chemical plants, power stations etc.) have been assessed - but for shorter times and using a database, at least on subsystem components, which has been broadened by practical experience. For systems where long term effects ought to have been assessed (e.g. agricultural developments, toxic waste disposal etc), no concerted effort has been made to achieve reliable predictive capabilities.

For the 10^4 - 10^6 y timescales which are considered in radioactive waste disposal, it is recognized that assessment of performance must be based on quantitative modelling of the system. In recent years many papers, including various overviews /1, 2, 3, 4, 5/, have been devoted to discussing the methodology and the applications of performance assessment for waste repositories. The ultimate rôle is assessing the safety of the final disposal facility:in practice, this will be done to provide input to the relevant safety documentation for obtaining regulatory licences for construction, operation and closure of a repository.

But performance assessments also play an important part in developments before this stage. They can be used for:

- developing concepts and checking their viability
- formulating site characterization programmes and guiding the course of the exploratory work
- aiding detailed design of disposal system components
- contributing to the technical/economic optimisation of the facilities

Today no country has yet, for geologic disposal of HLW or ILW, gone through all of the potential applications mentioned. Concept developments have been widely performed, usually accompanied by relatively crude assessments of expected performance. More extensive formal assessments of the basic feasibility of implementing safe disposal have been carried out recently in some countries e.g. Sweden, Switzerland, Germany /6, 7, 8/ and are planned soon in others, e.g. Canada, Finland. For selection of potential disposal sites, little formal use has been made of P.A. Most sites have been selected on other grounds backed up by qualitative judgements of the expected performance. In various programmes, however, concerted efforts are being made to ensure that the site characterization programmes currently underway receive maximum input from assessment studies. It is clear then that performance assessments have driven national waste disposal programmes to significant, but varying, degrees. For example, Canada, with less politically-imposed deadlines than many other countries, has consistently tried to guide programmes in this way /9/, whereas a recent USA review /10/ recommends increased integration of P.A. within the DOE programme.

2 P.A. METHODOLOGY TODAY

More work has certainly been invested so far into development of methods than into their application. International review documents and meetings, like the present one, have attempted to maintain a consensus on the most promising approaches. The earlier studies divided assessments into scenario and consequence analysis. It was recognized that scenario analysis is the more difficult issue: How can we be sure that our scenario set is complete? How can we assign probabilities to the diverse processes and events which can take place at the repository? Is it possible to adequately quantify the numerous data and their possible variations over long times? Many such questions are still open.

Consequence models involve mathematical description of physical and chemical processes. Model developments in many areas rapidly advanced to a stage where restrictions are now often due to data requirements rather than methods. This situation can in some instances be improved by further data collection (e.g. chemical thermodynamic data for modelling geochemistry; in other areas (e.g. discrete fracture flow modelling), it was recognized that a complete dataset will never be available so that other - stochastic - approaches are being developed. Even today, moreover, there are single processes or coupled effects which may be important in determining repository performance and which are not yet adequately modelled. One illustrative example is transport of water, dissolved species and gases through unsaturated clay buffer materials.

The separation of problem areas (scenarios, data, models) as described above has been often found to be less than ideal. Different approaches using stochastic models and data distributions to simulate future development of the disposal system are also being developed /11/. The evolution of the system is modelled taking account of the effects of slow, natural and repository induced processes and a statistical distribution of perturbing events is superimposed on this.

An increasingly important aim is more objective quantified assessment of **risks**, which should reduce the danger of basing judgements on particular predicted consequences of low probability. One obvious problem is that of combining uncertainties caused by incomplete knowledge, imprecise data and future changes.

3 FACTORS GOVERNING THE CHOICE OF P.A. METHODOLOGY

The fruits which the labours of all performance assessors must yield at intervals in repository project development are quantitative predictions of the future behaviour of a complete disposal system or of individual subsystems. These predictions must be sufficiently accurate and precise and should be accompanied by some indication of their inherent uncertainties. How does one choose the methology to be applied in such "milestone" assessments?

Three factors are important:

- the objective of the assessment
- the criteria for the assessment
- the timescales.

These factors, which explain the types of assessments which have been produced to date and will be produced in future, are discussed in turn below.

A Influence of P.A. aims on methodology

The aims of early assessments were to investigate basic feasibility of geologic disposal and also to identify important research areas. Particular examples of this latter application are the studies at NRPB where wide parameter variations were performed with simple, sequential, deterministic models /12, 13/ and work in Canada with the SYVAC code /14/ which can indicate the precision with which important modelling parameters need to be defined.

Various partial assessments have also been carried out in the progress of methodology development. Examples are the first generation of CEC studies involving the P.A.GIS methodology /15/, or more specifically Belgian assessments within this programme /16/. The reporting on these partial assessments often tends to de-emphasise the status of the final calculated results for risks or consequences with the justification that developments are incomplete or that sufficient site-specific data are not yet available.

A few performance assessments to date have had the objective of quantifying repository safety in as concrete a manner as possible for review and discussion within a more formal framework. First were the Swedish KBS 1 and KBS 3 projects /17, 6/ which were formally assessed as a prerequisite to future reactor operation. Similar work has recently been completed in Switzerland /7, 18, 19/. Because of the legalistic implications of such studies, maximum emphasis must be placed upon using methods and data which are transparent and auditable and have been verified and validated to the maximum possible extent. This can lead to a type of "bounding analysis" which is aimed not at the best estimates of reposiory performance but rather at reasonable upper bounds to expected consequences. For demonstrating that adequate safety is achievable, one may have to go no further than such approaches, as has been suggested for recent work in the USA /20/. The problems of proceeding to later project optimisation if only boundary-analyses have been performed are, however, obvious.

Finally recent performance assessments in the USA have been undertaken in the framework of the Environmental Impact Statements required for potential disposal sites in the USA. This is, of course, finally possible only when the results of site characterization programmes have become available. Preanalyses are, however, valuable to indicate problems in complying with suitability criteria. The types of problems perceived may then be directly related to these criteria, as discussed below.

B Influence of assessment criteria on P.A. methodology

Performance assessment approaches, methodology and data requirements are clearly dependent upon criteria to be applied. The bounding type analysis mentioned above (in particular for Sweden and Switzerland) have emphasised simple individual dose criteria. These allow flexibility in system design and analysis, including the recognition of the beneficial effects of dilution for nuclides with half-lives so long that significant geosphere transport might be possible. It can, however, lead to overemphasis on particular scenarios of small probability or of small total consequence (e.g. direct intrusion, concentrated outflow in a small water well serving few people).

In the USA, criteria are applied also to component parts within the total repository system (containment time, restricted near-field releases, minimum groundwater travel times). This is currently having a strong impact upon the efforts being invested in performance assessment. One result is that expert groups tend to concentrate on specific issues. This brings a useful concentration of expertise but tends to underemphasise integrated analyses and increases the risk of important issues being overlooked. For example a near-field release measured simply as the fraction of a given radionuclide released per year may contain insufficient data for proper geosphere transport analyses which require knowledge of physical and chemical forms of the radionuclides (valency, colloids, complexes etc.)

Finally, safety criteria may be deterministic or probabilistic and this will clearly influence the methodology applied. Swiss safety guidelines are (almost) completely deterministic at present; USA guidelines allow higher release rates for less probable scenarios; latest NEA suggestions /21/ are for proper risk rather than dose limits and safety authorities in some countries (e.g. GB, CDN) appear to be close to final adoption of such risk criteria. In general, there is a broad and growing awareness that uncertainties in far future predictions will always remain, and must be quantified as far as possible, and this is leading to a steady trend towards probabilistic, risk-based performance assessments. As will be discussed below, however, some problems remain to be cleared-up before full performance assessments of this nature can be handled in a formal regulatory process.

C Influence of timescales on P.A. methodology

Performance assessment methodology has been developing over the last years together with the development, and partly even the implementation, of disposal concepts and facilities. Assessments for guiding further work within a project are, in principle, required on a continuing basis and should be regularly updated. They will tend to use the methodology currently judged best even if formal verification and validation is not yet sufficient.

Figure 1 shows schematically the development of the "best" methodology with time following some form of classic learning curve. Validated methodology lags behind, is hopefully getting near, but will never quite reach the most advanced methods. When adequately validated models become able to provide the required quality of analyses we will reach a time T_C

- which is critical for modellers because further funding is difficult to justify! In the parallel case of reactor risk analyses this point is time has been postponed by progressively ratcheting up the project requirements! Currently we are in the time range (T_1 -T_c), that is methods are available for adequate analyses, but validation is incomplete.

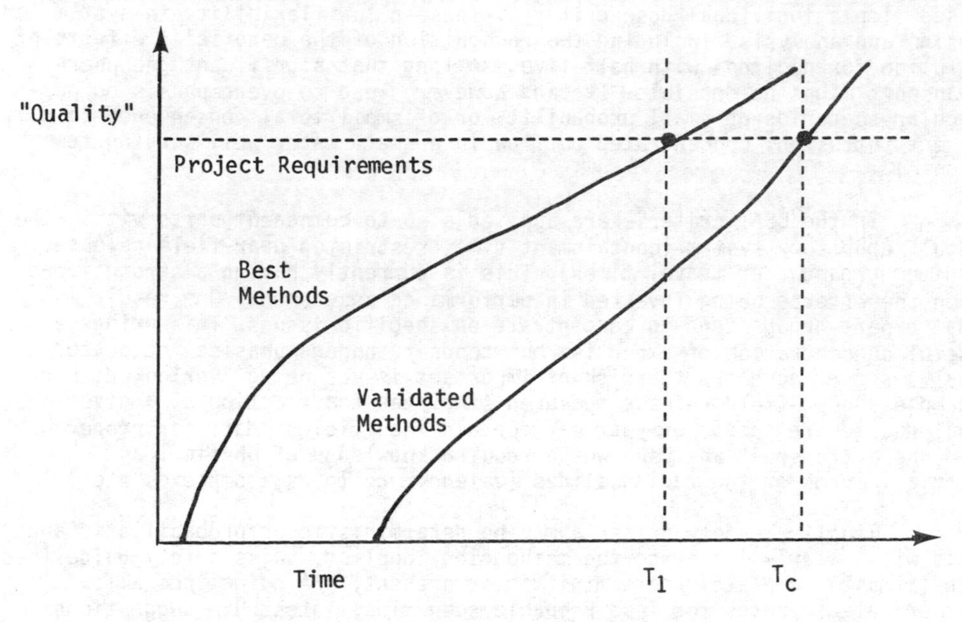

Figure 1: Development of P.A. Methodology with Time - Schematic

4 PRESENT STATUS OF PERFORMANCE ASSESSMENTS

Perusal of the proceedings of the many recent waste disposal meetings reveals innumerable reports on collecting data for safety assessments, very many papers on developing methodology, much less work on assessment of restricted subsystems and very few results of overall system analyses. One reason for this is the increasing need for a specific site at which full geological characterization has been performed.

For near-surface disposal of LLW, some countries (e.g. Britain, USA, France) already have sites in operation and specific analyses have thus been possible. For geologic repositories for ILW, site specific analyses have been possible in West Germany following site characterization /22, 23/, are in progress in Sweden along with complete site characterization /24/, and have begun in Switzerland where a specific potential disposal site has been identified and preliminary field data are available /25/.

For HLW no country has yet nominated and fully characterized a definitive disposal site. The Mol site in Belgium and the Gorleben site in Germany are the most advanced in this respect. Intensive work is beginning at the 3 potential USA sites, however, and with the foreseen timescales and funding levels these could be amongst the first HLW sites to be fully analysed, e.g. /26/. In Canada, a unique opportunity is presented by having a rock laboratory investigation site which (although not considered as a potential location) has the characteristics of a repository site and thus will allow an earlier exercise of a full site-specific performance assessment. Sweden is investigating a number of specific potential HLW disposal sites over the next years and will assess the performance of each individually. In Switzerland recent HLW assessments have been based on regional field studies and a specific site will not be characterized for several years.

The requirement for site-specific data and the fact that detailed final approaches will depend on the type of host rock and site has been recognized in international circles. But the fact that all final safety assessments will be subject to international comparison and review has also led to recognition that the greatest possible unity of approaches is desirable /4/. The following broad statements summarize the current consensus:

- The important rôle of P.A. in waste disposal planning and implementation is recognized
- Methods for modelling the most important processes determining repository performance are available (e.g. groundwater transport, corrosion, nuclide transport, heat transfer etc.)
- In some cases, data requirements restrict the applicability of the models (e.g. fracture flow)
- There are areas where research is needed to establish whether further modelling needs exist (e.g. coupled processes, colloid and complex behaviour)
- There are areas where, even though conservative models already allow bounding estimate analyses, better modelling is needed for optimisation (e.g. geochemistry, nuclide retention)
- There must be better validation of the existing performance assessment models, with emphasis on key safety issues (e.g. groundwater transport, matrix diffusion)
- The uncertainties associated with quantitative performance predictions must be more precisely determined
- The move to probabilistic methods will continue for the above reason, and also as a reflection of the inherent incompleteness in our knowledge of disposal system parameters over the distance and time scales to be considered.

Of all the points mentioned, the last two will be commented upon in more detail since they bring us to the front of performance assessment methodology as it is being developed today.

5 STATUS OF PROBABILISTIC METHODS IN P.A.

The difficulties which can arise when we try to assess the safety of any process by quantifying only its possible consequences are apparent. If we concentrate upon the most likely consequences, then we may pay insufficient attention to serious scenarios of low probability. If we emphasise maximum possible consequences, then we can arrive at a misleading perception of overall safety, as was the case following the earliest extreme estimates of maximum effects which could result from nuclear power plant accidents. To be as objective as possible we must quantify risks, i.e. for the case of repositories we must estimate probabilities of events which can initiate or influence nuclide release and we must assess also the probabilities of specific controlling parameters assuming certain values within their possible ranges.

Probabilistic approaches have been worked on for some time /27, 28, 29/. As mentioned earlier, however, few complete formal assessments have as yet been done probabilistically. Advances are being made, but problems remain. Most concrete progress has been in the probabilistic treatment of uncertainties in input parameters and the influence of these on predicted system performance; more problems remain in the treatment of uncertainties in scenaro definition.

Treatment of uncertainties resulting from possible variations in parameter values has been developed in Canada and in the CEC through use of codes (SYVAC, LISA) which allow data to be specified as probability distributions /14, 30/. Using different techniques, these input distributions are repeatedly sampled to provide data sets for successive system analyses. By performing numerous runs we can build up a plot of consequences with their appropriate relative frequences. For very simple analytical solution methods and simple analytical forms of probability distributions, it can also be possible to give results in probabilistic form without using Monte Carlo methods /31, 32/.

These probabilistic methods using parameter distributions rather than discrete values are valuable in several ways. They give a more realistic picture of the risks. By examining in detail runs contributing to the high consequence tail of the results curve, we can see which parameters most critically affect safety. Because the sensitivity to the sharpness of input parameter distributions is apparent, the analyses provide a valuable tool for guiding the expensive data collection exercises which are needed for site specific performance assessment.

But the methods must be used with prudence for several reasons. The code modules are of necessity simple; a lot of comparison with more complex models ("research models" in Canadian terminology) is needed and the final restrictions in the simple methods must not be forgotten. Careful use of statistics is needed to properly interpret the results, especially at the interesting high consequence end where fewer case histories are generated. Perhaps more important, however, is to guard against "over-interpretation" of the results of any set of runs. The relative frequencies of given consequences cannot be combined directly into a risk estimate since normally not all possible scenarios are covered. Quantifying the probability of fundamentally different scenarios can be more problematic.

The probabilities associated with future development of the repository system can be difficult to assess. Many possibly significant effects are of such low probability that statistical evidence is sparse (e.g. ice age effects, major erosion) and reliance must be on a proper mechanistic understanding of the processes involved. This can be difficult, in particular for some geologic scenarios which can have major consequences but have probabilities which are very low and difficult to assess. Examples are the appearance in future of major faults or the existence at present of such undetected faults.

It is thus clear that developments are still needed before full applications of probabilistic methods appear in formal performance assessments. Increased efforts are being made: first site specific analyses have been done for the Mol site; for the assessment of the potential HLW sites in the USA increased emphasis is being put upon quantification of uncertainties; Canadian work will lead to formal probabilistic assessments.

However, deterministic methods continue to have their place. Deterministic methods can be used for bounding analyses, which may in some cases be sufficient; they are needed to model the detailed physical and chemical behaviour of the disposal system and hence are important in validation process; they remain necessary for modelling complex systems. Currently and in the next years, therefore, a combination of determinatic and probabilistic methods will have to continue to be used in order to quantify as best possible our understanding of the evolution and possible future consequences of waste disposal systems.

6 CURRENT CHALLENGES FOR P.A.

The foregoing discussion leads to a list of aims for future work in performance assessment. These are summarized in Table 1. Most issues need no further comment. An exception is the point concerning documentation of assessments; this has not as yet been discussed in this paper, despite its high importance. Understanding the abstract modelling approaches often used in P.A. is not always simple for scientists of all disciplines, it is sometimes very difficult for decision makers (who must act on the results), and is often near-impossible for the interested public. These problems will increase as the methodology becomes yet more sophisticated. Therefore, significant effort must be devoted to transparent documentation of the approaches and the results.

Finally, the last item in Table 1, although already mentioned more than once, must be stressed as the most important task facing those involved in performance assessments. More complete, integrated analyses must be carried through. These reveal the weak points in our approach, build up valuable experience for the future formal licensing procedures and also add to the body of serious analyses which give an objective picture of the generally low risks associated with geologic waste disposal.

Table 1: Challenges for P.A.

- Improve **detailed process models** in some areas
- **Validate** the models more completely
- Improve **uncertainty** treatment - especially in scenario analyses
- Provide more input for **guiding field-work**
- **Complete more system assessments** giving **quantitative** predictions of expected performance

REFERENCES

1 IAEA, "Safety Assessment for the Underground Disposal of Radioactive Wastes"; Safety Series No. 56, IAEA, Vienna, 1981.

2 IAEA, "Concepts and Examples of Safety Analyses for Radioactive Waste Repositories in Continental Geological Formations"; Safety Series No. 58, IAEA, Vienna, 1983.

3 IAEA, "Performance Assessment for Underground Disposal Systems for Radioactive Waste"; Safety Series Report Category IV, IAEA, Vienna, 1984.

4 NEA, "Long-term management of high-level radioactive waste - The meaning of a demonstration"; OECD, Paris, 1983.

5 National Research Council, "A study of the isolation for geologic disposal of radioactive wastes"; National Academy Press, Washington D.C., 1983.

6 SKBF/KBS, "Final Storage of Spent Nuclear Fuel - KBS-3"; Stockholm, 1983.

7 Nagra, "Projekt Gewähr 1985"; Volumes 1 - 9, Baden/Switzerland, 1985.

8 PSE, "Zusammenfassender Abschlussbericht - Projekt Sicherheitsstudien Entsorgung"; Hahn-Meitner-Institut für Kernforschung, Berlin, 1985.

9 Wuschke D.M., Mehta K., Dormuth K.W., Andres T., Sherman G.R., Rosinger E.L.J., Goodwin B.W., Reid J.A.K., and Lyon R.B., "Environmental and Safety Assessment Studies for Nuclear Fuel Waste Management, Vol. 3: Post-Closure Assessment"; AECL TR 127-3, Pinawa, 1981.

10 Liebermann J.A. et al., "Performance Assessment National Review Group"; Report RFW-CRWM-85-01, Washington, 1985.

11 NEA, "Radionuclide release scenarios for geologic repositories"; Proc. of Workshop held in Paris, OECD/NEA, 1980.

12 Hill M.D., Grimwood P.H., "Preliminary Assessment of the Radiological Protection Aspects of Disposal of High-Level Waste in Geologic Formations"; NRPB-R 69, Harwell, 1978.

13 Hill M.D., "Analysis of the Effect of Variations in Parameter Values on the Predicted Radiological Consequences of Geological Disposal of High-Level Waste"; NRPB-R86, 1980.

14 Dormuth K.W, Sherman G.R., "SYVAC - A Computer Program for Assessment of Nuclear Fuel Waste Management Systems, Incorporating Parameter Variability"; AECL-6814, Atomic Energy of Canada Ltd., Chalk River, Ontario, 1981.

15 PAGIS, "Performance Assessment of Geological Isolation Systems - Summary Report of Phase 1: A Common Methodological Approach Based on European Data and Models"; EUR 9220 EN, Commission of the European Communities, Luxembourg, 1984.

16 Bertozzi G. and D'Aalessandro M., "A Probabilistic Approach to the Assessment of the Long-Term Risk Linked to the Disposal of Radioactive Waste in Geological Repositories"; Radioactive Waste Management and the Nuclear Fuel Cycle, Vol. 3(2), pp. 117-136, 1982.

17 KBS, "Handling of Spent Nuclear Fuel and Final Storage of Vitrified High-Level Reprocessing Waste"; Stockholm, 1978.

18 Issler, McCombie C., "Demonstration of the Feasibility of Safe Disposal of Radioactive Wastes: The Swiss Approach"; Paper presented at the International Meeting in Tucson/AZ, March 1985.

19 McCombie C., "Predicting the Safety Performance of a HWL Repository"; Paper presented at the International Topical Meeting in Pasco/WA, September 1985.

20 Pigford T.H. and Chambre P.L., "Reliable predictions of waste performance in a geologic repository; Int. Topical Meeting on High-Level Nuclear Waste Disposal, ANS, Pasco/WA, 1985.

21 NEA, "Long-term radiation protection objectives for radioactive waste disposal"; NEA, Paris, 1984.

22 PTB, "Zusammenfassender Zwischenbericht über bisherige Ergebnisse der Standortuntersuchung in Gorleben"; Physikalisch-Technische Bundesanstalt, Braunschweig, Mai, 1983.

23 GSF, "Andere Entsorgungstechniken, Abschlussbericht der Projekt-Phase 1, Technischer Teil"; KFK-AE Nr. 9, August 1982.

24 Hedman T. and Forsström H., "The Swedish final repository for low- and medium-level reactor waste"; Conf. Waste Management '84, Tucson/AZ, 1984.

25 Diebold P., Kappeler S., Sattel G., Tripet J.P., van Dorp F., "Definition of a site investigation programme for a potential Swiss low- and intermediate-level waste repository"; to be published in IAEA Colloquium of Site Selection, Design and Construction of Underground Repositories, Hannover, March 1986.

26 ONWI, "Performance Assessment Plan and Methods for the Salt Repository Project"; BMI/ONWI-545, Columbus, 1984.

27 Iman R.L., Helton J.C. and Campbell J.E., "Risk Methodology for Geologic Disposal of Radioactive Waste: SEnsitivity Analysis Techniques"; SAND78-0912, NUREG/CR-0394, Sandia National Laboratories, Albuquerque/NM, October 1978.

28 Cranwell R.M., Campbell J.E., Helton J.C., Iman R.L., Longsine D.E., Ortiz U.R., Runble J.E., Shortencarier W.J., "Risk Methodology for Geologic Disposal of Radioactive Waste, Final Report"; NUREG/CR-2573, Sandia National Laboratories, Albuquerque/NM, December 1982.

29 D'Alessandro M. and Bonne A., "Radioactive Waste Disposal into a Plastic Clay Formation: A Site-Specific Exercise of Probabilistic Assessment of Geological Containment"; Radioactive Waste Management, Vol. 2, Harwood Acad. Publ., Paris, 1981.

30 Saltelli A., Bertozzi G., Stanners D.A., "LISA - A Code for Safety Assessment in Nuclear Waste Disposal"; EUR 9306 EN, Commission of the European Communities, Luxembourg, 1984.

31 Pritzker A. and Gassmann J., "Application of simplified reliability methods for risk assessment of nuclear waste repositories"; Nuclear Technology, 1980.

32 Schaeffer D.L. and Hoffman F.O., Nucl. Technol. 45, 99, 1979.

SESSION II

THE LINKAGE BETWEEN PERFORMANCE ASSESSMENTS AND REGULATORY REQUIREMENTS

THE LINK BETWEEN GENERAL PERFORMANCE OBJECTIVES AND PERFORMANCE ASSESSMENTS OF DISPOSAL FACILITIES FOR RADIOACTIVE WASTE

Dr. P. Johnston
UK Department of the Environment

ABSTRACT

General radiation protection objectives for radioactive waste disposal have recently been proposed by an NEA expert group and are under discussion by the International Commission on Radiation Protection. Some national regulations concerning the safety of radioactive waste disposal facilities are currently formulated in terms of general objectives. This paper discusses the implications of setting general objectives for the conduct of performance assessments and for the presentation of the results of these assessments. Particular attention is given to regulatory developments in the European countries and Canada and to the procedures adopted within the CEC PAGIS project.

RELATION ENTRE LES OBJECTIFS GENERAUX DE PERFORMANCES ET L'EVALUATION DES PERFORMANCES DES INSTALLATIONS D'EVACUATION DE DECHETS RADIOACTIFS

RESUME

Un Groupe d'experts de l'AEN a récemment proposé des objectifs généraux de radioprotection applicables à l'évacuation des déchets radioactifs qui sont en train d'être examinés par la Commission internationale de protection radiologique. Certains règlements nationaux concernant la sûreté des installations d'évacuation de déchets radioactifs sont actuellement formulés en termes d'objectifs généraux. L'auteur analyse les conséquences de l'établissement d'objectifs généraux pour la réalisations des évaluations des performances et la présentation des résultats de ces évaluations. Il insiste particulièrement sur l'évolution de la réglementation dans les pays d'Europe et au Canada et sur les procédures adoptées dans le cadre du projet PAGIS de la CCE.

1. GENERAL RADIATION PROTECTION OBJECTIVES FOR RADIOACTIVE WASTE DISPOSAL

1.1 The rationale

The idea of establishing general radiation protection objectives for radioactive waste disposal is attractive from two different viewpoints:

- The major long-term concern about radioactive waste disposal is to restrict the radiological impact that might arise. It is therefore useful to have an indication in very general terms of what could be considered as an unacceptably high radiological impact. The approach of setting general limits and objectives is one that has been developed over many decades for radiation protection of the workforce at nuclear installations and for members of the public. The mechanism for establishing recommendations on such objectives already exists in the ICRP and in other international bodies.

- From the viewpoint of design of repositories for radioactive waste, it has been recognised for a number of years that the combination of several interrelated barriers to radionuclide movement can provide a basis for ensuring the long-term safety of a facility. However, the necessary complementarity of the barriers, such as the geological setting and engineered features of a disposal facility, makes difficult the generic specification of performance objectives for any one barrier in isolation. Specification of an overall performance objective therefore leaves open the possibility that a variety of repository designs in different locations could be consistently compared, even with different emphases on the role of each barrier to radionuclide movement

1.2 The proposals

General radiation protection objectives for current nuclear activities already exist; recommendations are given in ICRP publication 26 for protection of workers and members of the public. These recommendations are now embodied in most national regulations and already provide a basis for regulation of some radioactive waste disposal activities, notably the release of radioactive effluents. The recommendations in ICRP-26 apply to situations in which occupational radiation exposure and exposure to members of the public can be influenced by control of operating procedures and the source of radiation or radionuclides, as well as control, if necessary, of environmental transfer routes after accidental releases of radioactivity. However, the development of disposal facilities for radioactive waste must also take account of potential radiation exposures to members of the public far in the future. Institutional controls cannot be expected to be effective for more than a few hundred years, and it is therefore a major consideration in the design of disposal facilities for these wastes that continued control is not a necessity for safety. The safety assessments must be based on the assumption that control of the source itself or environmental transfers ceases to exist after a few hundred years. This means that radiation protection objectives for waste disposal must be understood as a set of targets in the planning, design and licensing of disposal system, with authorisation for disposal conditional on <u>predictive</u> radiological risk assessments being consistent with these targets.

It was recognised in 1982 that there was a need to define long-term radiation protection objectives that would acknowledge the eventual termination of institutional control, but that would be both consistent with current radiation protection recommendations and could be applied clearly and effectively in the design, construction and licensing of disposal systems. The OECD Nuclear Energy Agency set up an interdisciplinary group to suggest suitable objectives. A report, on which this paper draws, was published in 1984 [1]. The ICRP also established a task group to draft recommendations. This group has reported to the ICRP and the suggested recommendations are also reflected in this paper [2].

Predictive assessments will inevitably be subject to considerable uncertainity; not only uncertainity about the rates of processes that are expected to influence future radiological impacts, such as radionuclide transport in groundwaters, but also uncertainty about when some events and processes will occur, or even whether they will occur at all. There may be a very wide range of predicted annual radiation doses to members of the public, with the highest doses associated with very unlikely events or processes. In judging the acceptability of a waste disposal system, it is therefore necessary to consider not only the magnitude of annual exposures that might arise, but also the probabilities that various levels of annual dose will be received.

The protection of individuals can be ensured in two ways: One can define a risk limit to restrict the expectation value of annual dose from all exposure scenarios to an acceptable value; alternatively one can define a dose limit, or a set of dose limits, and also define acceptable probabilities of exceeding these limits.

1.2.1 The risk limitation approach

The NEA group and the ICRP task group have suggested limitation of risk as a primary objective in design of waste disposal facilities and in the authorisation of disposals. Risk is defined as the probability of a health effect leading to premature death of an individual or his immediate descendants. For some aspects of radioactive waste disposal, it has been convenient to further define risk as the product of three parameters:

H - the dose equivalent (the sum of the external dose in a year and the 50 year committed dose associated with that year's intake of radionuclides)

P - the probability that the dose will be received by an individual

r(H) - the probability of a health effect per unit dose equivalent (10^{-2} per Sv in the dose region associated with stachastic effects)

It was the intention of the NEA group that the risk limit should be a guarantee of protection to the individuals of the most exposed group. The dominant concern was to limit individual health risks. With this emphasis on individual protection, consistency can be retained with the ICRP system of dose limitation which expresses individual dose limits in terms of annual dose equivalents. The value of a risk limit can be related to the maximum annual dose that could be permitted have unit probability of being received: the annual dose limit.

For waste disposal systems, it may often be gradual processes that release and transport radionuclides to the environment, rather than discrete disruptive events. The presence of radionuclides in the environment may only arise thousands of years after the disruption or degradation processes initiating radionuclide release and transport. Furthermore, the presence of radionuclides in the environment and associated radiological consequences may persist for thousands of years. For these scenarios, it is more convenient to talk in terms of annual dose equivalents at the time radiation exposures occur, rather than to talk in terms of the dose commitment from a disruptive "event" at the time an "event" occurs.

A further consideration is that we are as much concerned with uncertainty in predicting impacts associated with gradual degradation and transport processes that will inevitably occur at some time, as we are with assessing the probabilities of unlikely disruptive events that could give rise to earlier or larger radiological impacts. Uncertainties in impacts of both gradual processes and probabilistic disruptive events can be handled in a consistent manner in relation to limitation of risks if risk is assessed as defined earlier.

The NEA group has suggested that national authorities should judge waste disposal practices against an individual risk limit for members of the public, which corresponds to the risk associated with current ICRP dose recommendations. It is suggested that a maximum risk objective shold be 10^{-5} per year, corresponding approximately to the objective of 1 mSv per year, as a lifetime average, recommended by the ICRP. It is further suggested that national authorities establish risk upper bounds, for particular waste disposal practices, that are only a fraction of the general limit.

The ICRP group has made suggestions along lines very similar to those of the NEA group. The numerical risk limit suggested by the ICRP group is identical to that suggested by the NEA group, but again it is stressed that this needs to be apportioned between different disposal facilities affecting the same group of individuals, and between today's practices and those in the future. The ICRP group has further suggested that when a continuous probability distribution of annual dose equivalents exists, reflecting uncertainities in modelling parameters or a variety of probabilistic events, the individual risk be defined as the integral of the risk associated with all potential annual dose equivalents that could arise from all events and processes affecting a waste disposal facility. This involves assessment of the probabilities of exceeding various levels of annual effective dose equivalent at various times in the future for individuals of the critical group. The risk to these individuals can be calculated by an integration of this function over all annual dose rates, as described in figure 2.

At a national level, the Department of the Environment in the UK has proposed principles for assessment of disposal facilities for low-and Intermediate-level waste [3] in line with the suggestions from these international bodies. To make allowance for exposure pathways not recognised at present, and to avoid prejudicing any future decisions that might lead to other activities that could cause radiation exposures to the same individuals that might be exposed as a result of radioactive waste disposal, a risk target for a single repository of one tenth of that suggested as the general risk limit has been proposed. A risk target of 10^{-6} per year is equivalent to an annual dose of 0.1 mSv in circumstances where doses are likely to arise with a probability near unity. This specification of a risk target is accompanied by a number of other requirements on siting, monitoring and documentation of disposal facilities and by a further general objective that future releases of radioactivity from a facility should not significantly increase the natural quantities of radioactive material in the general locality of the facility.

A very similar approach to regulation of waste disposal facilities is under discussion in Canada. In February 1985, a meeting of the Canadian Radiation Protection Association suggested that two basic criteria for acceptability of a disposal system be adopted [4]:

- The estimated risk to an individual during the post closure phase of disposal vault should not exceed an established risk level;

- The estimated probability of exceeding an established individual annual dose equivalent level should not exceed an established probability level

The first was seen as the primary criterion, and the second was proposed in order to address concerns about high annual dose levels even if the risk criterion is met. The second was seen to enable comparison to be made with natural background doses and current regulatory limits.

1.2.2 The dose limitation approach

Dose limitation is an established regulatory technique for ensuring adequate radiation protection of individuals affected by current nuclear activities. An identical approach for protection of members of the public who may be affected in the future by radioactive waste disposal activities is attractive at least from the point of view of consistency with current regulatory practice and simplicity. No new concepts are involved, and a dose limitation approach has been provisionally adopted in several countries:

Regulations entitled "Safety criteria for disposal of radioactive wastes in a Mine" were published in the FRG Federal register in 1983 [5]. These contain a general objective that the disposal of radioactive wastes must guarantee the protection of man and the environment from harm caused by ionizing radiation. This is reinforced by a specific requirement that for all reasonable scenarios of radionuclide release from a repository individuals should not receive doses of greater than 0.3 mSv per year

The Swiss nuclear safety authorities issued general guidelines for radioactive waste disposal in 1980 [6]. These contain the following restrictions on individual doses:

- Radionuclides which, as a result of realistic events and processes, return from a repository to the biosphere, must at no time give rise to an individual dose exceeding 0.1 mSv per year;

- If the effects of several repositories overlap, the sum of the doses to an individual from all repositories should not exceed 0.1 mSv per year.

In Sweden, no official radiation protection objectives for radioactive waste disposal exist, but the following criteria were used in the KBS-3 assessment [7]:

- The expected doses to individuals in the most highly exposed group should be less than 0.1 mSv per year.

- Even under unfavourable circumstances, individual doses should be less than 1 mSv per year.

- The repository should not essentially alter the natural radiation environment in the repository surroundings.

All the "dose limits" in these regulations and suggested criteria are associated with qualifications about the reasonableness, the reality or the likelihood of the release scenarios leading to the exposures. The uncertainty about the magnitude and timing of releases and their radiological impact requires that any attempt to establish a dose limit must be accompanied at least by a qualitative statement about the probability of exceeding the limit.

1.3 Consistency between risk and dose limitation

The essential difference between risk limitation and dose limitation is simply that in risk limitation a quantification is needed of the probabilities of an individual receiving various levels of dose, whereas in dose limitation a qualitative statement about the probability of various release scenarios may be sufficient. The two approaches can be consistent, in that they can assure a similar degree of radiation protection to the individuals most at risk. Equivalent risk and annual dose values (10^{-6} and 0.1 mSv per year) appear to be emerging as the basis for regulation. The degree of consistency in the level of protection will however also depend on the way in which compliance with limits is demonstrated and on the interpretation of the concept of "reasonableness" associated with the choice of scenarios for dose limitation.

2. THE DEMONSTRATION OF COMPLIANCE WITH GENERAL PERFORMANCE OBJECTIVES

It has been recognised [6] that a "demonstration" of the safety of radioactive waste disposal must be indirect and must involve a convincing evaluation of the disposal system performance on the basis of predictive modelling confirmed by a body of varied technical and scientific data [8].

The first pre-requiste for a demonstration of compliance with general performance objectives is the ability to assess the variation from year to year of annual individual dose rates, even far in the future, given a description of a disposal facility, its hydrogeological setting if on land, or its oceanographic setting if in the marine environment. This paper only discusses assessments of land-based disposal facilities. There has been considerable progress in recent years in computer-based predictive modelling: Events and processes by which radionuclides could re-enter the surface environment even from deep-mined repositories have been assessed and the most important in terms of their probability or radiological impact have been simulated in models of the processes involved.

Uncertainties will always exist in the parameter values used in models to describe release and transport of radionuclides. The way in which parameter values are chosen and uncertainties are handled will depend on whether compliance with a dose limit or a risk limit is to be demonstrated. The CEC PAGIS project is faced with the problem of conducting safety assessments in such a way that compliance with both risk and dose limits can be demonstrated. Reference is made in the following Sections to the procedures adopted for the PAGIS project [9]. Demonstration of compliance with a dose limit is conceptually familiar and this is deal with first:

2.1 Compliance with dose limits

If demonstration of compliance with dose limitation is the goal, it may be adequate to select a limited number of scenarios and sets of parameter values for detailed asessment.

In the CEC PAGIS project, both "normal" or "altered evolution" scenarios are assessed, and the "reasonableness" of both scenarios and of parameter values is given consideration in their selection. For each release scenario, a "best estimate" calculation is made using realistic values of parameters in the models to generate a curve of annual dose to an individual of a critical group as a function of time. For the PAGIS assessments, it has been agreed to define a hypothetical individual for each exposure pathway who is representative of the group which would receive the highest dose via that pathway. The individual is taken to be an adult with constant diet and metabolism unless it is obvious that children or infants would receive higher doses via the pathway.

A dose limit can be deemed to be respected if the calculated annual individual dose at all times, and for all scenarios, is less than the limit. The situation is simple if all selected scenarios are associated with individual annual doses within an adopted limit. However, if there are scenarios, or possible sets of modelling parameters, associated with higher individual doses, a judgement has to be made about their "reasonableness". It is this judgement that is the key element in the consistency of dose and risk limitation when equivalent limits are adopted. If "reasonable" is interpreted in the sence of "possible", even though of low probability, a dose limit may be much more restrictive than an equivalent risk limit. If "reasonable" is interpreted as "likely", a dose limit may be less restrictive than an equivalent risk limit.

2.2 Compliance with risk limits

The quantification of probabilities as well as annual doses is essential in risk assessment and in demonstration of compliance with risk limits. Two stages are possible in an evolutionary development from dose assessment to risk assessment.

For "normal" and "altered evolution" scenarios of radionuclide release, it might be thought that risk could be calculated simply by multiplying the "best estimate" annual doses by an estimate of the probability of the scenario and by the probability of a fatal health effect per unit dose equivalent. For a number of reasons that are explained later, this procedure may underestimate risks in the context of risk limitation suggested by the ICRP.

A more rigorous risk assessment must take account of uncertainties in modelling parameters. In risk assessment, it is the expectation value of the annual dose, rather than the most probable value, that must be multiplied by the probability of a fatal health effect per unit dose equivalent. Parameter sampling techniques such as those used in SYVAC and LISA codes can generate curves of the expectation value of dose as a function of time, and therefore of individual risk as a function of time. The risk is a single number that can be compared with whatever target or limit is adopted, and the risk limit would be deemed to be respected if the assessed risk is less than the limit at all times. Sampling techniques are applicable to both "normal" and "altered evolution" scenarios, and are used in Canadian and UK assessments as well as in parts of the CEC PAGIS project.

There are disadvantages as well as advantages in a sampling approach. Rather than estimating a single "most likely" value for each modelling parameter of importance, it is now necessary to estimate the range of possible values and the shape of the probability density function for each parameter. For example, rather than adopting the mean value of permeability from a set of measurement in a geological formation, it is necessary to characterise the distribution of permeability measurements in terms of its shape (log-normal, normal etc), its width (the standard deviation for a normal distribution), and its mean. This is undoubtedly difficult and does add considerably to the burden of data acquisition. The need to assess the radiological impact associated with a range of different parameter values also increases the computational times. The advantage is that ranges of parameter values better reflect our knowledge of natural systems.

In many circumstances, the delayed and gradual transport of radionuclides may be perturbed by natural events or human actions that may or may not occur, but that could give rise to earlier or larger radiological impacts. In some situations, these events which alter the evolution of a disposal system can be described in a probabilistic way, and can give the dominant contribution to the risk from a disposal facility. For example, seismic and tectonic phenomena, which could modify groundwater flow, are important for disposal in some geological formations, and future human activities such as drilling, mineral exploitation and mining could have direct and indirect influences on some repositories. If radioactive waste is disposed of near the surface, other natural events and human actions may have a potential disruptive influence and may give rise to immediate or delayed radiological impacts. Risk assessment requires that the probability of exceeding various annual doses as a result of such events must be estimated. This means that the frequency of occurrence or the annual probability of such events needs to be estimated. Estimation of probabilities remains one of the more uncertain parts of many assessments.

3. OUTSTANDING ISSUES IN THE LINKAGE BETWEEN GENERAL RISK LIMITS AND PERFORMANCE ASSESSMENTS

Two issues appear to be of continuing concern in this important linkage. One is related to the practical methods of calculating risk and the other to the assurance of completeness in risk assessment. The two issues are discussed separately below:

3.1 Practical methods of calculating risk

A distinction must be made between scenarios of delayed or gradual radionuclide release leading to long-term contamination of the environment, but low-levels of radiation exposure, and low-probability scenarios of human intrustion or sudden disruption of a disposal facility leading to higher levels of short-term radiation exposure.

For the first class of scenarios, of individual risk as defined in the NEA report on long-term objectives [1] can best be related to the product of three parameters as indicated in section 1.2.1. This product involves the probability that, at the time of interest, an individual will be exposed to a particular annual dose.

This probability may be related to the cumulative probability that the environmental contamination will have been produced at the time of interest by prior radionuclide releases from a disposal facility (Figure 1). The probability may equally be related to the uncertainties on modelling parameters through the distribution of possible values (Figure 2). Whichever of these situations exists, a large degree of agreement has now developed about the way risks can be assessed.

It has proved more difficult to develop a consistent and compatible method of calculating risk for the second class of scenarios associated with short-term exposures for which the individuals at risk may only be exposed for a few weeks or a few years: There are disadvantages in trying to associate annual doses with such scenarios. A resolution of this issue has become particularly pressing in the UK for risk assessments of near-surface disposal facilities for which intrustion and disruption scenarios can give the dominant contribution to individual risk. It is suggested that the individual risk for this class of scenarios could be assessed as for operating reactors by taking the product of three parameters:

H - The dose (both external and from radionuclide uptake) associated with the intrusion or disruption "event", integrated over the period for which it is received within an individual's lifetime;

P - The annual probability of instigation of the "event";

r(H) - The probability of a fatal health effect per unit dose equivalent appropriate for the dose rate and duration of exposure.

An illustration of procedure for risk assessment as specified above is given in figure 3.

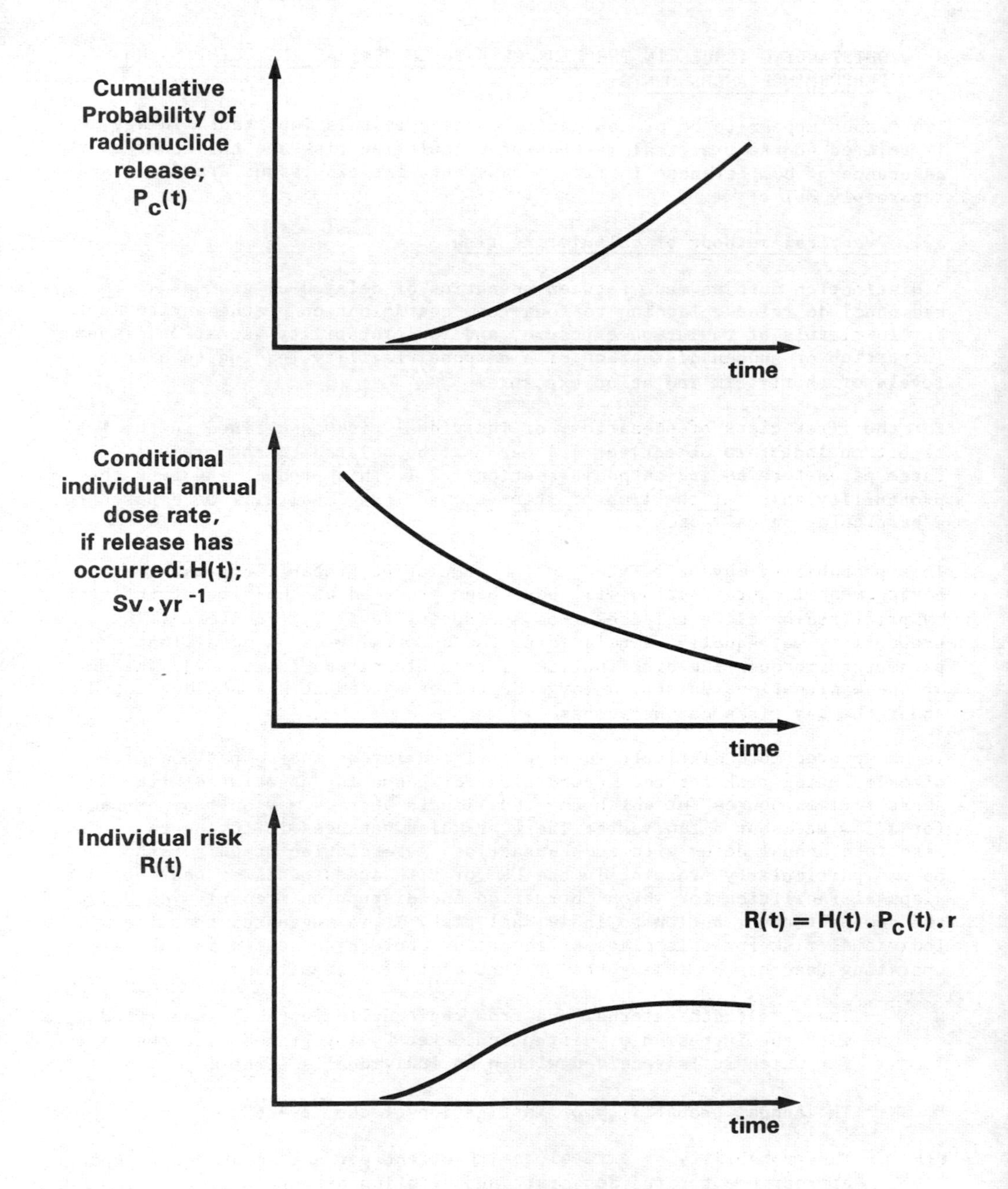

Figure 1:

An illustration of the procedure of risk assessment for probabilistic scenarios leading to long-term environmental contamination.

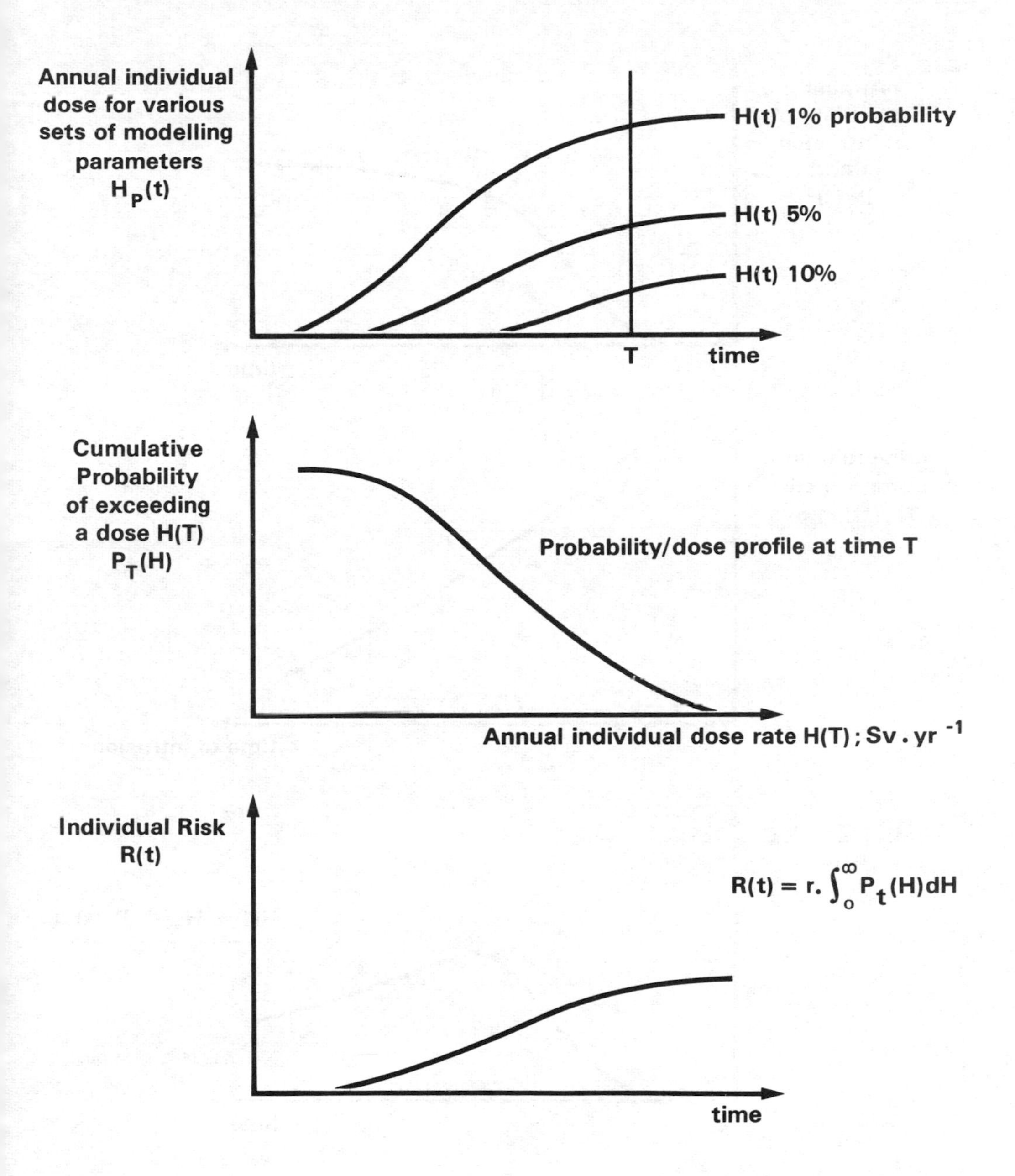

Figure 2:

An illustration of the procedure of risk assessment for uncertain parameter values represented by probability density distributions.

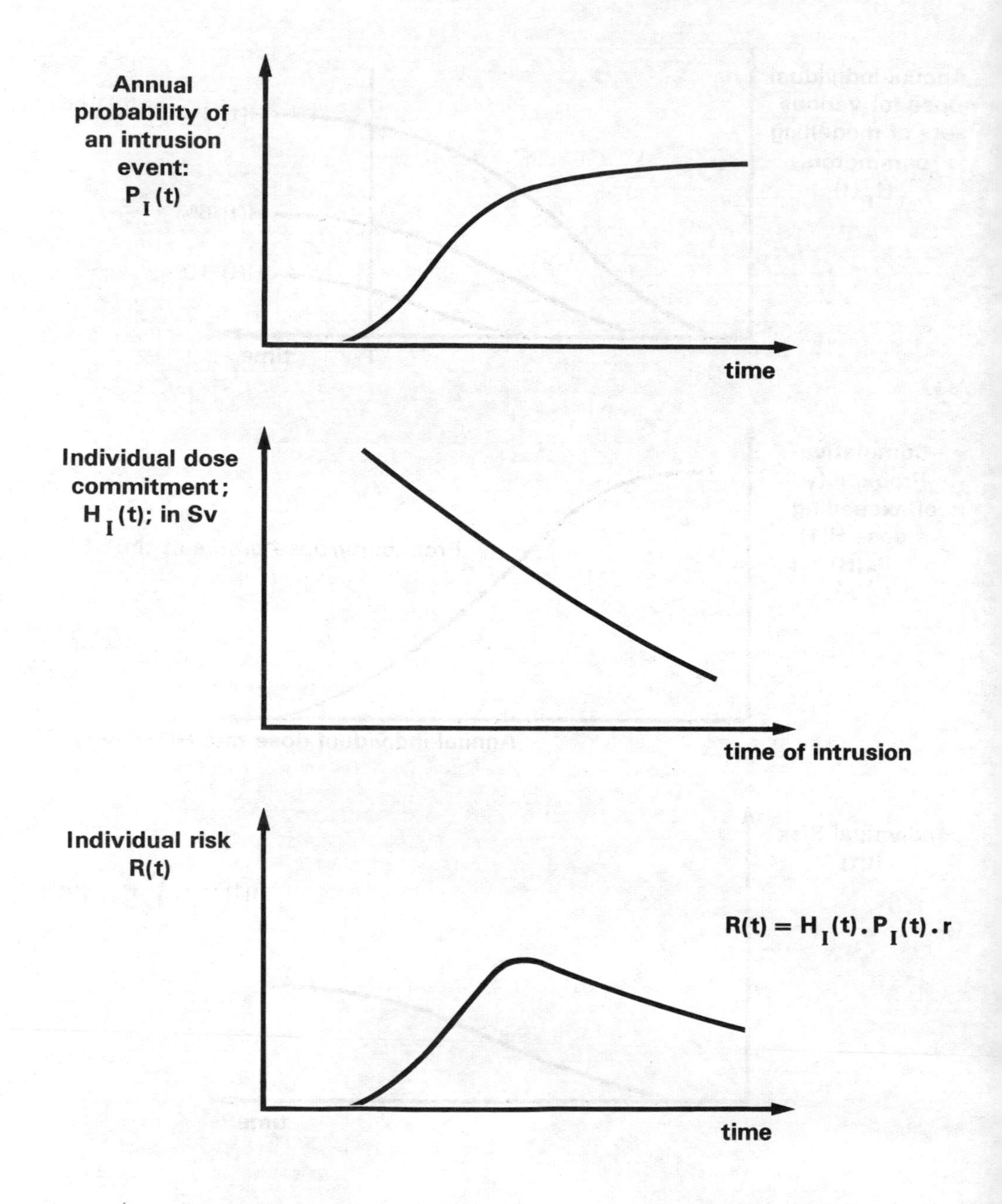

Figure 3:

An illustration of a procedure of risk assessment for probabilistic intrusion scenarios leading to short-term individual exposures.

3.2 The assurance of completeness in risk assessment

It will always be difficult to ensure in a safety assessment that all major contributions to risk are given due consideration. The problem is two-fold;

3.2.1 Scenario identification

It has been stressed by the ICRP that all important release, disruption and intrusion scenarios must be identified. In some circumstances, and for some repository concepts, the problem appears simple; for example, a deep repository in granite is unlikely to be disrupted by surface phenomena, and attention can be focused on gradual degradation and radionuclides transport processes, possibly enhanced by glaciation, seismic and tectonic phenomena. For other disposal facility concepts, it is less easy to restrict the number of processes and scenarios pertinent to a comprehensive assessment. Recent studies of near-surface disposal have identified over a dozen scenarios which could contribute significantly to individual risk [10]. These range from gradual dispersion of radioactivity, gas generation and emanation, to erosion and disruption leading to surface contamination and direct, but inadvertant, intrustion during civil engineering work. When such a diversity of potential exposure scenarios exists, there can be no absolute guarantee that all important contributors to risk have been assessed. The assurance of completeness must derive from a progressive screening out of unimportant contributors to risk from a longer list of scenarios than is eventually retained for detailed assessment. The fact that the contribution to risk from a variety of scenarios has been assessed, and that some contributions are shown to be negligible, can then contribute to confidence that no major contributions to risk have been ignored.

The need for an assessment of a wide variety of scenarios for radiation exposures has been recently highlighted by studies in the UK of the risks associated with intermediate-level waste in a mined disposal facility at a depth of a few hundred metres. A great deal of attention was given to scenarios of slow radionuclide release and transport in groundwater, showing risks of the order of 10^{-10} to 10^{-8} per year. A relatively simple assessment of risks associated with inadvertent drilling through such a repository then indicated risks of the order of 10^{-8} to 10^{-6} per year [11]. While such results represent little more than an initial screening of scenarios, they do help to indicate which types of exposure scenario merit further detailed analysis.

3.2.2 Selection and sampling of modelling parameters

Even when scenarios which are major contributors to overall risk have been identified, a second problem is to ensure that the risk is adequately reflected by the selection of parameter values used in modelling radiological impacts. This is a critical issue in the comparability of risk estimates from so-called "best estimate" assessments and from parameter sampling assessment methodologies. A more detailed discussion of the problem is given below, but recent assessments have indicated that in many waste disposal situations the major contribution to risk may be associated with low probability combinations of parameter values, well on the pessimistic side of "best estimates". This situation arises when the probability of an individual receiving an annual dose in particular dose range falls off more slowly than 1/H at annual dose values above the "best estimate" values. An assurance of completeness in risk assessment can only then be achieved by an extension of parameter selection or sampling well into the pessimistic side of all parameter probability distribution functions until it is observed that incremental contributions to risk decrease significantly.

To illustrate the problem, reference can be made to three different characteristic annual doses related to a probability distribution of predicted annual dose: The "most probable" annual dose, Hp; the median dose, H_m, and the expectation value of dose, H_e. The most probable dose is the dose at the peak of a probability distribution function. There is a 50% probability of exceeding a median dose, and it can be orders of magnitude higher than the "most probable" dose. The expectation value of dose is given by the integral $\int_0^\infty H \cdot P(H)\, dH$. For a distribution function with a long tail in the high dose region, it can be significantly greater than the median dose.

There are significant differences between the three different characteristic annual doses in actual assessments conducted by the DoE for a nominal disposal facility in clay [12]. These are illustrated in Figure 4. Much of the difference can be attributed to the observation that relative contributions to risk increase as a function of individual dose well into the high-dose tail of the probability density functions of annual dose. Evaluations using SYVAC to assess a repository in clay have shown that the most probable annual dose is near-zero; no radioactivity is calculated to reach the biosphere. Clearly in this situation an assessment of the most probable dose will give a zero result although there is a finite risk associated with the full range of possible parameter values.

It is possible to envisage assessment procedures that will give consistent estimates of total risk, with a reasonable assurance that all major contributions to risk are covered. Procedures can be envisaged both for assessments using detailed models and for assessments using simple models and parameter models.

A possible procedure using detailed models might be as follows;

- Make an estimate of annual dose as a function of time with "median" or "most probable" modelling parameters

- Use this as the basis for a sensitivity analysis to determine the extent to which key parameters need to change in order to produce a significant increase in maximum annual dose

- Assess the changed probability of a new set of key parameters.

By an iterative process of sensitivity analysis and changes in modelling parameters, curves could then be traced of the cummulative probability of exceeding various levels of annual individual dose at specified times. The iterative tracing of these curves could be continued until the point is reached when the incremental contributions to risk in additional dose intervals decreases significantly. There would then be a reasonable assurance that all major contributions to risk had been included in the total risk derived from the curve of cummulative probability of exceeding various levels of annual dose [figure 2].

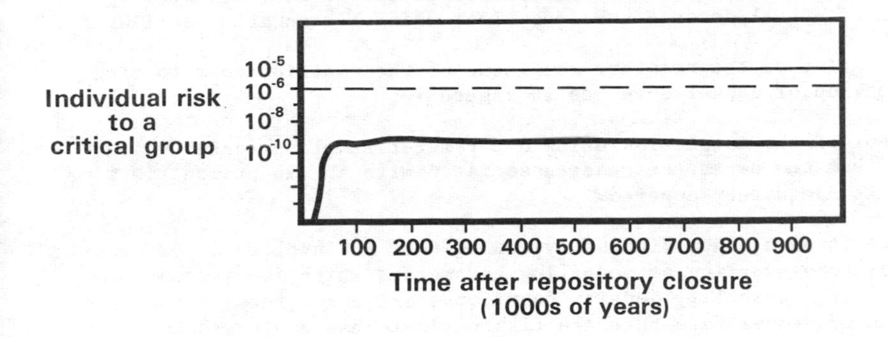

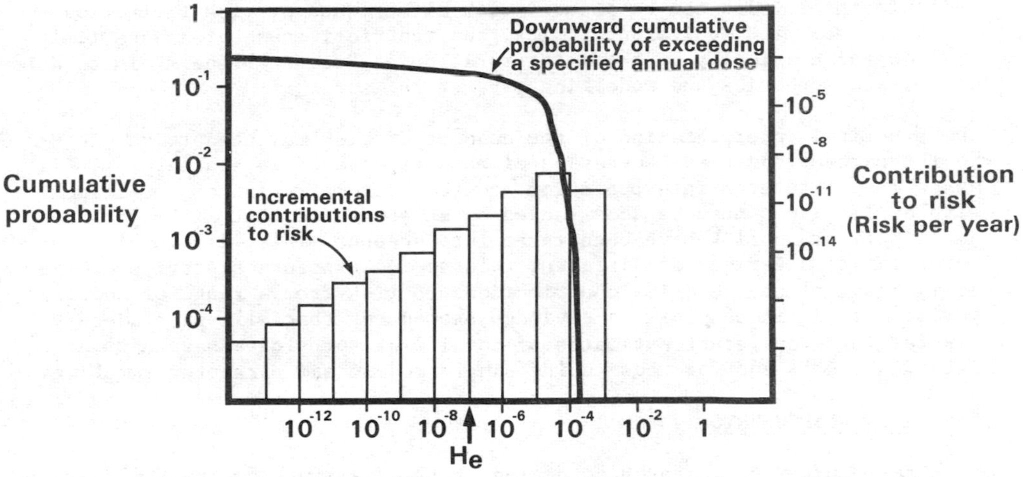

Figure 4:
Risks and annual doses associated with I-129 release from a repository for ILW in clay: UK DoE/SYVAC

For parameter sampling techniques a possible procedure might be as follows

- A preliminary identification of the shape of the probability density function of annual doses at various times using Monte-Carlo sampling;

- Identification of the relative magnitude of the contributions to risk as a function of annual dose (as in figure 4);

- Assessment of the total risk using a parameter sampling scheme which ensures that the parameter space associated with the major contributons to risk is adequately covered.

This procedure is being investigated by the UK DOE. It involves a system of relative importance sampling of modelling parameters which focuses on the high-dose tail of probability density functions, and a progressive increase in sample size to demonstrate that the risk estimate has converged to a stable value.

4. CONCLUSIONS

There is attraction in establishing general radiation protection objectives for radioactive waste disposal both from a regulatory and repository design viewpoints. Some appropriate general objectives have already been suggested at national and international level. These may take the form of dose limits for reasonable scenarios of radiation exposure, or of risk limits which reflect both doses and their probabilities of occurance. A similar degree of radiation protection can be assured with either dose or risk limitation, although even with equivalent limits, the restrictiveness of a dose limit will depend critically on the interpretation of "reasonableness" in relation to release scenarios and modelling parameters.

The practical interpretation of the concept of risk may need to be adapted to the circumstances and scenarios of most importance in a safety assessment. Whatever interpretation adopted, demonstration of compliance with a risk limit must be accompanied by an assurance that all major contributions to risk have been taken into account. This will involve a screening of a variety of different release and exposure scenarios and an exploration of the relative contributions to risk from a range of modelling parameters. It is possible to envisage procedures that will give adequately complete and consistent estimates of total risk both for assessments using detailed models and for those using simple models and parameter sampling.

5. ACKNOWLEDGEMENTS

This paper draws on research conducted in the framework of the CEC PAGIS project and on research for the UK DoE. Contributions and advise from Mr Cadelli, CEC; Ms M D Hill, NRPB; Dr B Thompson DoE; and Dr R Lyon, AECL are gratefully acknowledged. The views expressed in this paper do not necessarily represent UK Government policy.

6. REFERENCES

[1] Long term radiation protection objectives for radioactive waste disposal; OECD 1984.

[2] The application of radiation protection principles to radioactive waste disposal; ICRP to be published, 1985.

[3] Disposal facilities on land for low- and intermediate-level radioactive waste: principles for the protection of the Human Environment, DoE, UK 1983.

[4] Proceedings of the CRPA workshop on Radioactive Waste Management criteria, Toronto, Canada (1985).

[5] Safety criteria for disposal of radioactive waste in a Mine; FRG Federal register; 1983.

[6] Protection goals for the Disposal of radioactive wastes in Switzerland; U. Niederer, proc. Symposium on Waste Management, Tucson, USA, vol 1 (1981)53.

[7] Final storage of spent nuclear fuel: KBS-3; vol IV, Swedish Nuclear Fuel Supply Co: Stockholm 1983.

[8] Long-term management of high-level radioactive waste: The meaning of a demonstration. OECD/NEA 1983.

[9] PAGIS workshop on methods of estimating risks from high-level waste disposal and presenting results; summary of discussions and conclusions, CEC 1985.

[10] Dames & Moore report.

[11] Calculations of the radiological impact of disposal of unit activity of selected radionuclides: G M Smith, H S Fearn, C E Delow, G Lawson and J Davis, DoE/RW/85-096.

[12] An initial examination of a procedure for the post-closure radiological risk assessment of underground disposal facilities for radioactive waste: B G J Thompson and T W Broyd, UK DOE Disposal Assessment Technical Report TR-DOE-7 1985.

A PRESCRIPTIVE APPROACH - THE LINK BETWEEN THE PERFORMANCE ASSESSMENTS OF THE U.S. DOE PROJECTS AND REGULATORY REQUIREMENTS OF NRC AND EPA

A. E. Van Luik
Pacific Northwest Laboratory
Richland, Washington

N. A. Eisenberg, D. H. Alexander
U.S. Department of Energy
Washington, D.C.

ABSTRACT

The U.S. regulatory approach to high-level radioactive waste and spent nuclear fuel disposal has been called 'prescriptive'. This is because standards, requirements, and criteria that the implementing agency must follow, have been prescribed by the regulatory agencies, and are largely based on conceptual research and preliminary performance assessments performed by or for these agencies. This approach, and its resulting regulations, reflects the U.S. institutional structure, and does not differ greatly from the regulatory goals and approaches being used within the OECD community of nations. The linkages between U.S. performance assessment strategies and features of the U.S. high-level waste and spent fuel disposal regulations are briefly explored.

RESUME

Un exemple de méthode prescriptive : Le lien entre l'évaluation des performances des projets du Département de l'énergie des Etats-Unis et les exigences réglementaires de la "NRC" et de l'"EPA".

On a qualifié de méthode prescriptive, l'approche réglementaire adoptée aux Etats-Unis pour l'évacuation des déchets de haute activité et des combustibles irradiés, car les normes, exigences et critères dont il convient d'assurer l'application, sont prescrits par les organismes de réglementation. Ces prescriptions sont fondées largement sur des recherches théoriques et des évaluations de performances préliminaires effectuées par ou pour ces organismes de réglementation. Cette méthode et les réglementations qui en résultent reflètent la structure du gouvernement des Etats-Unis, et ne diffèrent guère des buts et des méthodes des pays de l'OCDE. Cet exposé présente brièvement les liens entre les stratégies d'évaluation des performances des systèmes d'évacuation aux Etats-Unis et les caractéristiques des réglementations pour l'évacuation des combustibles irradiés et des déchets aux Etats-Unis.

1. INTRODUCTION

The United States and a number of the other member nations of the Organization for Economic Cooperation and Development (OECD) are currently researching ways to dispose of high-level radioactive waste and spent nuclear fuel. The goal of the high-level radioactive waste and spent nuclear fuel disposal programs in the OECD-member nations is qualitatively identical: to assure acceptably small dose commitments to present and future generations. However, between nations there may be quantitative differences in how this goal is defined (for example, differences in the quantitative definition of minimal dose commitments and the definition of the time period encompassed by regulations).

It is perceived that the U.S. approach to regulating its high-level radioactive waste and spent nuclear fuel disposal is unique within the OECD community. The United States is perceived to have adopted a 'prescriptive' approach, wherein regulatory institutions prescribe certain limits below which a given repository must perform in order to be licensed. This approach is sometimes regarded as different from approaches in Sweden, Germany, Switzerland and other nations, where implementing agencies have typically been charged with proving the safety of mined geologic repository disposal. If preliminary safety evaluations are accepted by the governing authority, detailed standards and criteria will then be developed to assure that the actual repository will function as well as indicated in these preliminary evaluations.

Another difference generally perceived to exist between the U.S. 'prescriptive' approach and the approaches being used by other OECD-member nations concerns the U.S. regulatory emphasis on the functioning of the site and the engineered barrier system, with the performance standard being a prescribed limit on cumulative radionuclide releases past a boundary defined within the geosphere. By contrast, other nations generally focus on resulting health effects as their performance standard.

These perceived differences are, however, reflections of differences in procedural and institutional approaches more than actual differences in technical approach. The U.S. regulatory institutions, the U.S. Environmental Protection Agency (EPA) and the U.S. Nuclear Regulatory Commission (NRC), have both evaluated implications of the EPA standard regarding health effects [1,2,3]. The EPA standard is based on an upper limit allowance of ten health effects (latent cancer fatalities) per 10,000 years per 1,000 metric tons of heavy metal of reprocessed or unreprocessed spent nuclear fuel in a repository [3]. In its standard, the EPA defined this allowable health-effects level and then prescribed an upper limit to cumulative radionuclide releases to a specific boundary within the geosphere such that this health-effects level will not be exceeded. By so doing, the EPA effectively removed biosphere transport, dose-to-humans, and health-effects modeling and its inherent issues and controversies from the licensing process.

The regulatory approach of the NRC exists to give enhanced assurance that a licensed repository will perform in accordance with the EPA standard. The NRC's emphasis on the site and engineered barriers is, therefore, part of a conservative national institutional approach to providing 'reasonable assurance' that the EPA standard will not be violated. This conservative institutional approach is mandated by the Nuclear Waste Policy Act of 1982 (NWPA) (Public Law 97-425 [4]).

2. REQUIREMENTS, STANDARDS, AND CRITERIA

A. The Nuclear Waste Policy Act of 1982 (NWPA)

Since the 1950's, the United States has had a commercial nuclear energy program under a variety of agencies that are predecessors to the U.S. Department of Energy (DOE). These agencies were charged with both regulatory and research and development responsibilities until the 1970's. As a result of legislation in 1970 that reorganized government environmental activities, the EPA was given the authority to establish generally applicable environmental standards for radioactivity. In 1974, a reorganization of government nuclear energy activities resulted in the creation of the NRC and its regulatory role, thereby separating the nuclear regulatory functions from the research and development functions of the U.S. government. The DOE was created in 1977.

The NWPA established "a definite federal policy" regarding the disposal of high-level radioactive waste and spent nuclear fuel. The NRC, EPA, and DOE are to continue their roles as mandated by previous legislation, but with respect to the siting and development of nuclear waste repositories, the NWPA focuses and defines their respective efforts.

In the United States, congressional mandates such as the NWPA are usually broad statements of policy that are implemented through the issuance of regulations by regulatory agencies. These regulations, when issued as final rules, have the force of law.

The purpose of the NWPA is in part "to provide for the development of repositories for the disposal of high-level radioactive waste and spent nuclear fuel" and "to establish a program of research, development, and demonstration regarding the disposal of high-level radioactive waste and spent nuclear fuel." This law and the federal regulations that implement it suggest that high-level radioactive waste (reprocessed spent nuclear fuel) and non-reprocessed spent nuclear fuel be disposed of in a mined, geologic disposal system (repository). The evaluation of disposal options other than deep geologic repositories is also required, however. The NWPA specifies that requirements, standards, and criteria be written to govern the permanent disposal of high-level radioactive waste and spent nuclear fuel so as to "provide a reasonable assurance that the public and the environment will be adequately protected."

The NWPA specifically mandates that the U.S. DOE, the NRC, and the EPA produce certain planning and regulatory documents to impact and help define the U.S. performance assessment effort. The DOE is to produce a "Mission Plan" and general guidelines for site selection. The Mission Plan has recently been published [5], and the general siting guidelines were issued as Chapter 10 of the Code of Federal Regulations, Part 960 (10 CFR 960) [6]. Similarly, the EPA was to produce standards applicable to a mined geologic disposal system for high-level radioactive waste and spent nuclear fuel, which were also recently issued as a final rule as part of Chapter 40 of the Code of Federal Regulations, Part 191 (40 CFR 191) [3]. The NRC's applicable regulations, consisting of technical requirements and criteria, are contained in the Code of Federal Regulations, Chapter 10, Part 60 [7].

The NWPA specifies that the NRC is to license DOE's repository. The licensing procedure outlined in the NWPA requires DOE to submit a license application to the NRC allowing DOE to build, operate, and close a mined geologic disposal system for high-level radioactive waste and spent nuclear fuel. This license application and its supporting documents will contain the performance assessments from which the NRC is to judge whether or not there is reasonable assurance that applicable regulatory standards, requirements, and criteria are met.

Prior to submitting a license application, however, the DOE must follow a complex site-selection procedure [4,6]. The procedure includes writing an environmental assessment for each site nominated as suitable for characterization; a site characterization plan for each site nominated for characterization; and an environmental impact statement for the site recommended from among those characterized. Public hearings, and consultation and cooperation with the states are mandated as part of the decision-making process. The final site-selection authority is apportioned between the president, the congress, the state suggested for the location of the repository, and the affected Indian tribe on whose reservation the site is located. The latter two governmental entities may disapprove a president's recommendation to congress. The congress may then either resolve to approve the president's recommendation, thereby ending the site-selection process, or not resolve to approve the president's recommendation, thereby continuing the site-selection process.

B. The Mission Plan for the Civilian Radioactive Waste Management Program

Features of the Mission Plan [5] that are salient to this discussion are a firm commitment to the permanent disposal of spent nuclear fuel and the plan to consolidate fuel rods prior to placement in a disposal canister (removing the individual fuel rods from assemblies and close-packing the rods in canisters).

The Mission Plan [5] recognizes that reprocessing spent nuclear fuel is presently not an economic option in the United States; however, if this fuel is reprocessed in the future, the resulting high-level wastes will be accommodated by the repository program. Presently, it is assumed that consolidated spent fuel will be the major waste form in the first geologic repository. Therefore, U.S. performance assessment activities must include a program of spent fuel research designed to characterize the spent fuel, and to provide sufficient data so the behavior of the spent fuel may be predicted within certain tolerances with reasonable assurance.

C. The EPA Standard

According to EPA's 40 CFR 191 standard [3], the engineered and natural barrier systems, constituting the postclosure mined geologic disposal system, may not release radionuclides to the "accessible environment" in excess of the amounts allowed by Table 1 of 40 CFR 191. Table 1, reproduced here, specifies radionuclide-specific cumulative curie release limits for 10,000 years after permanent closure of the disposal system. The "accessible environment" is defined as the land surface and that part of the lithosphere outside of a "controlled area" that is less than 5 km in any direction from the outer boundary of the original location of the radioactive waste in the disposal system. These release limits are based on considerations of the results of environmental transport and population-dosimetry modeling. The standard, therefore, is based on a consideration of population dose commitments and resulting health effects.

The EPA 40 CFR 191 standard makes specific demands regarding the performance assessment methodology that DOE is to employ to provide a "reasonable expectation" that this standard will be met. The language of the standard best conveys the probabilistic performance assessment approach that is mandated:

"191.13 Containment Requirements

(a) Disposal systems for spent nuclear fuel or high-level or transuranic radioactive wastes shall be designed to provide a reasonable expectation, based upon performance assessments, that the cumulative releases of radionuclides to the accessible environment for 10,000 years after disposal from all significant processes and events that may affect the disposal system shall:

(1) have a likelihood of less than one chance in 10 of exceeding the quantities calculated according to Table 1 (Appendix A); and

(2) have a likelihood of less than one chance in 1,000 of exceeding ten times the quantities calculated according to Table 1 (Appendix A).

(b) Performance assessments need not provide complete assurance that the requirements of 191.13(a) will be met. Because of the long time period involved and the nature of the events and processes of interest, there will inevitably be substantial uncertainties in projecting disposal system performance. Proof of the future performance of a disposal system is not to be had in the ordinary sense of the word in situations that deal with much shorter time frames. Instead, what is required is a reasonable expectation, on the basis of the record before the implementing agency, that compliance with 191.13(a) will be achieved." [3]

D. The NRC's Requirements and Criteria

The NRC sought to provide increased confidence that the 40 CFR 191 standard will be met by promulgating a number of technical requirements and criteria specific to the site and to the engineered barrier system in its 10 CFR 60 regulation [7]. The site requirement is that the ground-water

TABLE 1. RELEASE LIMITS FOR CONTAINMENT REQUIREMENTS AS GIVEN IN APPENDIX A, SUBPART B, 40 CFR 191 [3]

(Cumulative Releases to the Accessible Environment for 10,000 Years After Disposal)

Radionuclide	Release Limit per 1000 MTHM or Other Unit of Waste (curies)
Americium-241 or -243	100
Carbon-14	100
Cesium-135 or -137	1,000
Iodine-129	100
Neptunium-237	100
Plutonium-238, -239, -240, or -242	100
Radium-226	100
Strontium-90	1,000
Technetium-99	10,000
Thorium-230 or -232	10
Tin-126	1,000
Uranium-233, -234, -235, -236, or -238	100
Any other alpha-emitting radionuclide with a half-life greater than 20 years	100
Any other radionuclide with a half-life greater than 20 years that does not emit alpha particles	1,000

travel time before waste emplacement be sufficiently long to assure that most of the gamma-emitting radionuclides dominating the total activity of the repository during the thermal period will not reach the accessible environment. The engineered barrier requirements are that the waste packages provide substantially complete containment of radionuclides while fission-product decay-heat and radiation are high, and that the engineered barrier system allow only small fractional radionuclide releases over long periods of time after substantially complete containment has been lost.

The criteria that are related to these requirements prescribe a preemplacement 1,000-year travel time to the accessible environment boundary (i.e., <5 km) and give a 300- to 1,000-year range for the containment period that may be set for a given mined geologic disposal system. In addition, after the period of substantially complete containment, releases are not to exceed one part in 100,000 per year of the inventory of any radionuclide calculated to be present at 1,000 years after permanent closure. An exception exists for radionuclides released at a yearly rate that is less than 10^{-8} of the total 1,000-year system curie inventory. For the latter radionuclides, the annual release rate limit is the ratio of 10^{-8} of the total 1,000-year system curie inventory, to that radionuclide's inventory at 1,000 years. This somewhat complex exception allows slightly higher releases for lower inventory radionuclides.

These criteria specifically apply only to the case where no unanticipated processes or events are assumed. Unanticipated processes and events, according to NRC's 10 CFR 60 regulation [7], are "those processes and events affecting the geologic setting that are judged not to be reasonably likely to occur during the period the intended performance objective must be achieved, but which are nevertheless sufficiently credible to warrant consideration". The determination of the likelihoods of unanticipated processes and events and of resulting releases is highly site-specific. Until these determinations have been made, it will not be known whether or not unanticipated process or event scenarios may lead to significant likelihoods of unacceptable releases for a given site.

E. The Engineered Barrier System and the Multiple Barrier System Requirement

The NWPA [4] requires that NRC promulgate technical requirements and criteria that "shall provide for the use of a system of multiple barriers." The NRC, accordingly, has defined a geologic repository to include "the portion of the geologic setting that provides isolation of the radioactive waste," and an "engineering barrier system" (10 CFR 60) [7]. In effect, the definitions given in 10 CFR 60 describe a mined geologic disposal system as being composed of a natural barrier system and an engineered barrier system, thereby constituting a multiple barrier system.

The NRC and EPA emphasize that the engineered barrier system itself is also to be a multiple barrier system, however. EPA's 40 CFR 191 [3] suggests that a canister, a waste form, and materials placed over or around a waste form may each be barriers. NRC's 10 CFR 60 [7] defines the engineered barrier system so that, from an isolation perspective, its main functional component is the waste package. The waste package, in turn, is the waste form, and its container(s), shielding, and any absorbent or other materials that may be packed around each individual container.

In the regulatory context, there must be some justification for assuming that a given waste package component is actually a barrier. Both 10 CFR 60 and 40 CFR 191 define a barrier in terms of its ability to prevent or substantially delay the movement of radionuclides. Crediting a given component with barrier properties will, therefore, require some evaluations of its expected performance.

F. Aspects of a Typical Spent Fuel Waste Package System's Performance That Need to be Assessed

The spent fuel waste package may consist of a stainless steel, low-carbon steel, or other metal container containing spent fuel rods [5]. These rods will likely have been consolidated, or removed from the assemblies that kept individual fuel rods separated for use in reactor cores [5]. Each rod is clad in a hull either made of Zircaloy, an alloy that is about 98% zirconium, or stainless steel [8]. It is likely that a very small percentage of these cladding hulls may have some small defects [9,10], which could allow some ground-water contact with irradiated UO_2 shortly after canister failure. However, this does not detract from the fact that clad spent fuel is an important barrier within the engineered barrier system, and its behavior under repository conditions must be understood in order to predict the long-term performance of the engineered barrier system in a performance assessment.

Establishing the behavior of the spent fuel waste form under repository conditions is complicated by the fact that irradiated UO_2 is not the hard ceramic, relatively homogeneous material that it was prior to irradiation. Irradiated UO_2 tends to be friable and somewhat fractured along grain boundaries, depending on its power history. It is possible that UO_2 oxidation, as well as the migration of more volatile fission products under high temperatures, may contribute to the formation of intergranular gaps that could contain volatile radionuclides [11]. In addition, grain-boundary phase radionuclides may be more susceptible to dissolution by ground water than grain-interior phase radionuclides. Thus, a three-phase radionuclide release behavior may be expected, and has been reported for the spent fuel waste form [12,13]. These three phases appear to be: 1) a rapid release rate, upon loss of containment, for volatiles located in gaps; 2) a relatively high short-term release rate for grain-boundary phase radionuclides; and 3) a slow, long-term release rate for radionuclides that are part of the irradiated UO_2 matrix within grains. This behavior has implications for the performance modeling of the spent fuel waste form [13].

About 20% of the mass inside a typical spent fuel waste package may be Zircaloy [14]. A considerable amount of stainless steel may be present in some spent fuel waste packages since about 5% of U.S. spent fuel is clad in stainless steel [10]. Therefore, these metallic cladding materials and their corrosion products may also be barriers within the engineered barrier system. Whether or not cladding is a barrier in the regulatory sense depends on the behavior that can be established for it and its corrosion products under repository conditions, especially any nuclide-specific sorbent properties. Research needs to establish these properties before defensible performance assessment modeling can be done that incorporates the effects of this possible barrier.

By design, the metal container is a barrier during the containment period, and after a container has been breached, its corrosion products may persist in the environment near the container surface. These materials may also continue to constitute a barrier to the migration of specific radionuclides.

Finally, packing materials are to be used in some repositories as barriers to control ground-water flux and to selectively delay the migration of some radionuclides by sorption. Packing materials may consist of bentonite clay mixed with crushed host rock [5,15,16].

The NRC's release rate requirement [7] applies to the engineered barrier system, and the waste package is the most important part of that system. Therefore, to show compliance, performance assessments of spent fuel waste packages in a repository need to consider the entire waste package as a system consisting of a container, cladding, waste form, and, for some repository designs, packing materials.

3. THE U.S. DOE PERFORMANCE ASSESSMENT STRATEGY

The DOE is developing a performance assessment strategy, with the cooperation of four DOE projects, to show compliance with the standards, requirements, and criteria developed by the EPA and NRC. The four projects are responsible for repository siting studies at potential basalt, crystalline, salt, and tuff host rock sites.

The diversity of host rock types being considered for repository sites in the U.S. program is a significant difference between the United States and most other individual OECD-member nations [17], and dictates that site-specific aspects of the U.S. program will require the development of unique and generic modeling approaches to allow realistic simulations of geologic repository systems and their subsystems. It is assumed that sensitivity and uncertainty analyses will be used as part of the overall performance assessment process to determine the importance of perceived information needs, the degree of accuracy required in data, and the degree of certainty provided by a given model. It is recognized that limitations in either data or models may require the use of conservative bounding assumptions.

The DOE performance assessment strategy presently consists of the components shown in Figure 1. Because both the EPA standard [3] and the NRC regulation [7] emphasize that probabilistic evaluations must be a part of the process of showing compliance, scenario definition and evaluation is parallel to the calculation of expected performance. This is symbolic of their equal importance to the regulatory compliance calculations, which require that all significant scenarios, likely or unlikely, expected or unexpected, be evaluated. (A scenario is any process or event, natural or anthropogenic, that could alter the disposal system and result in radionuclide release to the accessible environment.)

It is also important to realize that in the United States the licensing process is a legal process, according to the NWPA [4]. Both site selection and licensing are part of a well-defined public decision-making process, with technical input. The technical input consists, in part, of performance assessment regulatory compliance calculations. Therefore, peer review of all facets of technical input to the decision-making process is necessary to assure its quality. Technical peer review judgments must be auditably documented in terms of the methodologies applied and the thoughts and procedures used by the experts to arrive at judgments or decisions.

A. An Illustrative Approach to Compliance Calculation

For illustrative purposes, a simplified repository system model could be conceived that consists of a source-term model and a contaminant-transport model. The source-term model could be used to estimate the expected range of waste package lifetimes and radionuclide release rates from the engineered barrier system. These release rates may then be used as input to the contaminant-transport model, which simulates radionuclide transport from the engineered barriers and provides estimates of cumulative releases to the accessible environment. This simplified system model may now be used to illustrate an approach to calculating compliance with the EPA standard [3] and the NRC criteria [7].

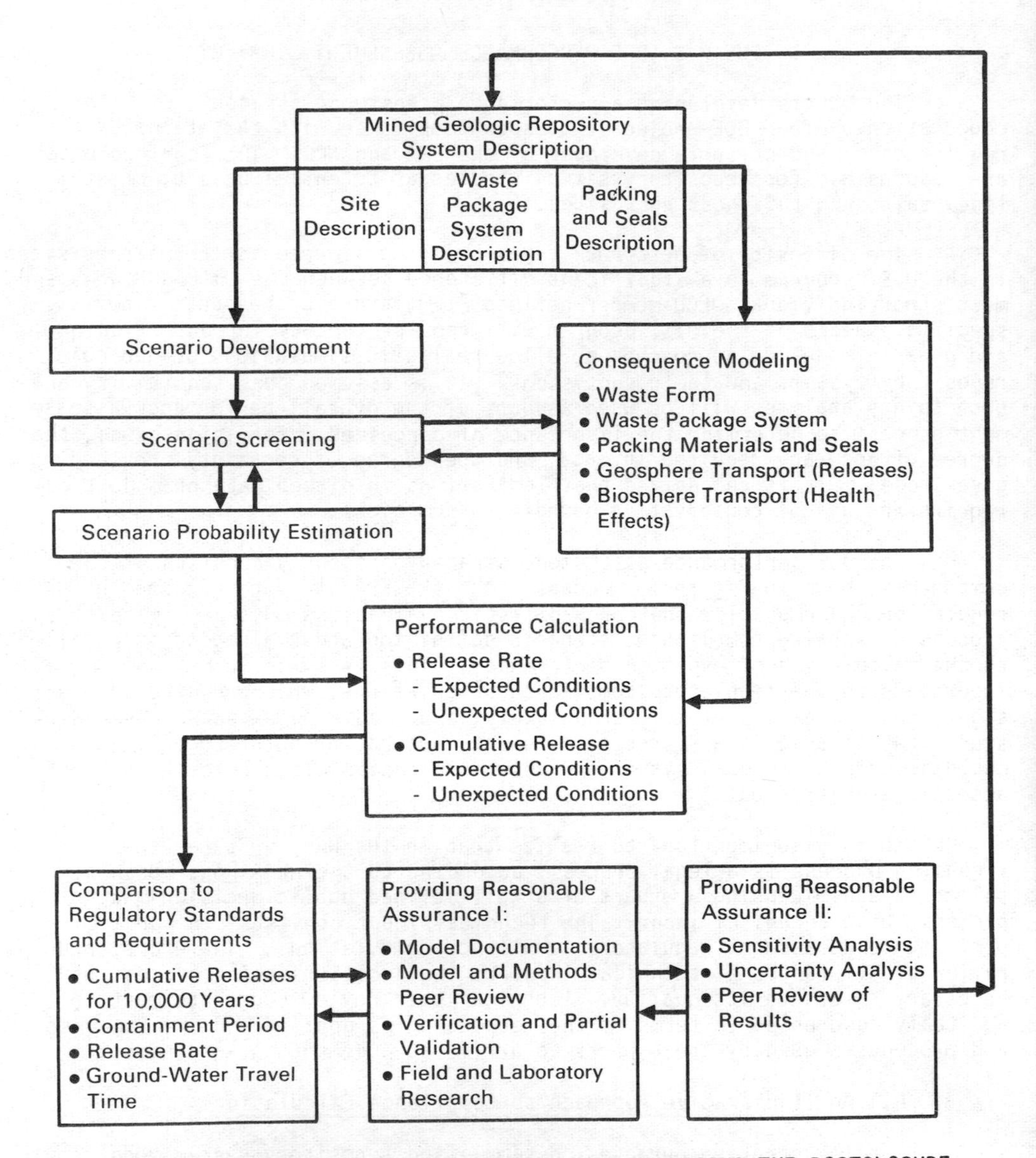

FIGURE 1. COMPONENTS OF THE STRATEGY FOR ASSESSING THE POSTCLOSURE PERFORMANCE OF THE GEOLOGIC REPOSITORY SYSTEM

B. Compliance with the EPA Standard

Output from the contaminant-transport model may be configured as a cumulative complementary probability distribution function (CCDF), which is required to show compliance with the EPA standard [3]. Figure 2 illustrates

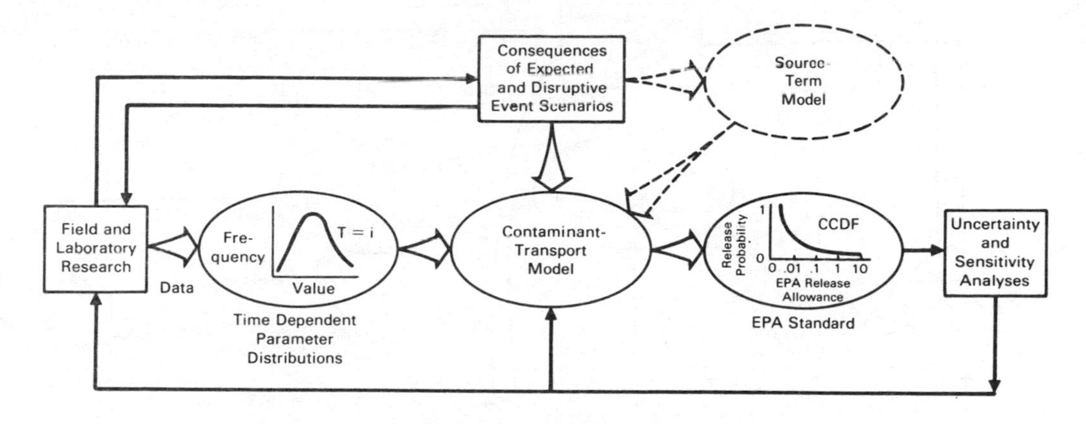

FIGURE 2. POSSIBLE STRATEGY FOR ADDRESSING THE EPA STANDARD

this plausible approach by showing the inputs (broad arrows) to the compliance calculation to be 1) stochastic and probabilistic input data; 2) information on scenarios; 3) radionuclide transfer information from the source-term model; and 4) the processing of this information by the contaminant-transport model. Sensitivity and uncertainty analyses follow to see if "reasonable assurance" has been achieved, and results feed back (line arrows) to the models and input data. The compliance calculation is, therefore, perceived to be an iterative process that may result in model modifications and the identification of additional data needs before the calculation is satisfactorily completed. This illustration is simplistic since its purpose is only to show main linkages between a variety of research and modeling activities needed to demonstrate compliance with the EPA standard [3]. Satisfying the EPA standard is going to be a complex, multifaceted problem [18].

C. Compliance with NRC Criteria

Showing compliance with the NRC's prewaste-emplacement ground-water travel time requirement [7] involves a judgment of the adequacy of the site hydrologic data base and the site conceptual hydrologic model supported by that data base. The travel time requirement requires a concensus of expert opinion, since the adequacy of data and the technical defensibility of its modeling and interpretation are essentially expert judgments. Documenting these judgments to describe each consideration with an auditable series of facts and informed, but perhaps informal, criteria is no trivial task, and requires formal methodologies and procedures. Figure 3 illustrates that showing compliance with the ground-water travel time requirement is dependent on the hydrologic conceptual model.

Returning to the simplified repository system model assumed previously, its source-term model might be used to show compliance with the NRC's engineered barrier system criteria [7] according to the scheme shown in Figure 3. As in Figure 2, inputs to the calculation are shown as broad arrows, and feedback loops, showing the iterative nature of the compliance calculation process, are shown as line arrows.

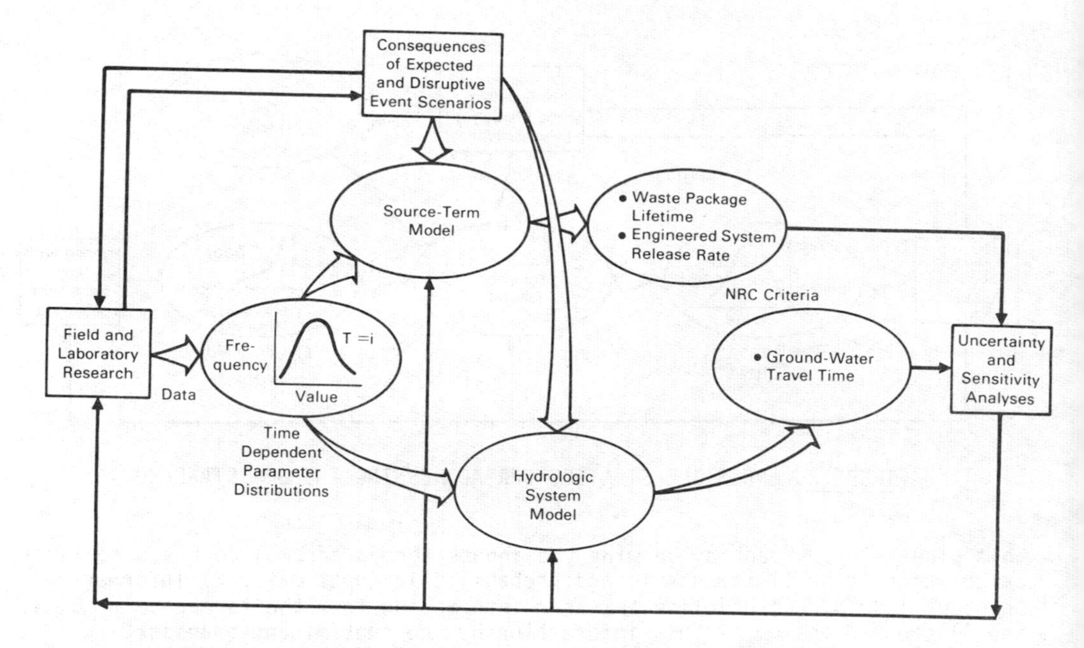

FIGURE 3. POSSIBLE STRATEGY FOR ADDRESSING THE NRC CRITERIA

The source-term model for releases from the engineered barrier system may also be used to assess the performance of the waste package components--the waste form (clad spent fuel), the container, and the overpack and/or packing material (if any). The actual engineered barrier source-term model should represent or bound the behavior of all relevant components and structures, their interactions, and the internal and external processes that affect performance. Variations in the approaches being developed for site- and design-specific assessments reflect the host rock specific design diversity within the U.S. program.

Principles being considered in formulating the DOE project performance assessment strategies may, in part, be gleaned from project publications [19,20]. It has been observed that in the United States, some preliminary system performance assessments have employed unrealistic bounding conditions to ensure conservative results [21]. With respect to the engineered barrier system source term, these bounding conditions have usually assumed 1) release from bare spent fuel, taking no credit for metal cladding and the container, or 2) instantaneous failure of the metal barriers from each waste package in the repository at 300 or 1,000 years, the minimum containment time range that may be required by the NRC's 10 CFR 60. These approaches have been criticized because neither approach allows the degree of conservatism to be established relative to a real case, and they are not capable of providing significant guidance for in situ testing or design development [21].

A conceptual model of the source term should consider radionuclide release from the boundary of a realistic engineered barrier configuration, avoiding physically impossible and worst-case scenarios [21]. In reality, water must penetrate the canister and the cladding before it can reach a significant amount of the spent fuel. After the penetration of the metal barriers, the waste must be dissolved from a maze of perhaps 1,000 closely packed fuel rods before it can pass through corrosion-induced apertures in the canisters. The release of most radionuclides from the waste package is controlled by the solubility of the UO_2 matrix, as well as chemical interactions with the surfaces of barriers and barrier alteration products that will line exit paths.

By accounting for enough of these processes and events, the source-term model may be used to establish a sufficiently realistic but conservative loss-of-containment time and radionuclide release rate. Processes within the waste package should be identified and characterized for assessing the behavior of the engineered barrier system and its components as a function of time, design, and repository conditions. Waste package monitoring and testing under simulated repository conditions, up to the time of permanent closure, should provide an understanding of these processes, and permit limited validation of the codes used to model the source term [21].

4. THE PRESCRIPTIVE APPROACH IN THE OECD CONTEXT

There are a number of apparent differences between the regulatory requirements that affect the U.S. repository program and the national repository programs of other OECD-member nations. These differences are largely the result of institutional and scale differences, reflecting the existing roles of regulatory institutions in the United States and the multirepository, multihost-rock site-selection process that is part of the U.S. program. Controlling the exposure risk to present and future generations is the common goal of programs in the United States and other OECD-member nations [22,23]. Moreover, OECD-member nations, including the United States, are pursuing this common goal through multiple-barrier, defense in-depth approaches [5,24-29].

Apparent technical differences between the United States and other OECD-member nations include: 1) the requirement to show cumulative releases to the accessible environment rather than doses to individuals or populations [3,23]; 2) the 10,000 year regulatory period, compared with an undefined, and therefore indefinite period in other countries [3,24,30-32]; and 3) the engineered barrier release rate limit, which as yet finds no parallel in the regulation of other nations [7]. The containment period requirement of 300 to 1,000 years [7] is not dissimilar, at 1,000 years, from preliminary goals being discussed in some national programs [24,29,33], but is much less than goals apparently being pursued by others [34]. It is not yet possible to determine if present, preliminary goals are to become regulations for these programs [34-37].

These differences in regulatory approaches mean that the U.S. performance assessment strategy must focus on the engineered barrier system as an entity in and of itself, which creates a notable difference between the performance assessment strategy of the United States as compared with the total-system approaches of other nations. However, this focus represents an institutional approach to providing added assurance that the primary goal, reducing risk to an acceptable level, can be met.

It is apparent that no nation can successfully demonstrate regulatory compliance by modeling alone. A rigorous, comprehensive, defensible data base must exist to allow model development and at least partial validation. The U.S. program may, on the whole, be more focused on the engineered barrier system than are its counterparts in other nations. However, the development of data bases on the characteristics and behaviors of clad spent-fuels and high-level wastes, on metal barriers and packing and sealing materials, and on the behavior and potential transport mechanisms for radionuclides in the host rock, provides fertile ground for meaningful work of mutual interest.

The radionuclide transport modeling needs for some U.S. performance assessment programs have obvious parallels to the needs of those countries considering either salt or granite as potential host rocks. Wherever parallel needs exist, whether in data or modeling, cooperation has the potential for being mutually beneficial in terms of saving precious national resources. The United States wishes to cooperate wherever such mutual benefit may be identifiable.

This work was supported by the U.S. Department of Energy under Contract DE-AC06-76RLO 1830.

REFERENCES

1. C. B. Smith, D. E. Egan, Jr., W. A. Williams, J. M. Gruhlke, C-Y Hung and B. L. Serini. 1982. "Population Risks From Disposal of High-Level Radioactive Wastes in Geologic Repositories," EPA 520/3-80-006 Draft Report, U.S. Environmental Protection Agency, Washington, D.C.

2. N. R. Ortiz and K. K. Wahi. 1983. "Technical Assistance for Regulatory Development: Review and Evaluation of the EPA Standard 40CFR191 for Disposal of High-Level Waste. Executive Summary." NUREG/CR-3235; SAND82-1557 (6 Vols.) Sandia National Laboratories, Albuquerque, New Mexico.

3. EPA. 1985. "U.S. Environmental Protection Agency, 40 CFR Part 191, Environmental Standards for the Management and Disposal of Spent Nuclear Fuel, High-Level and Transuranic Radioactive Wastes." Fed. Reg., Vol. 50, pp. 38066-38089 (September 19, 1985), Final Rule, Washington, D.C.

4. "Nuclear Waste Policy Act of 1982." Public Law 97-425, 42 USC 10101 through 10226, January 7, 1983.

5. DOE. 1985. "Mission Plan for the Civilian Radioactive Waste Management Program." DOE/RW-0005, Vol. 1 of 3 volumes, Office of Civilian Radioactive Waste Management, U.S. Department of Energy, Washington, D.C.

6. DOE. 1984. "U.S. Department of Energy, 10 CFR Part 960, Nuclear Waste Policy Act of 1982; General Guidelines for the Recommendation of Sites for the Nuclear Waste Repositories." Final Siting Guideline, Fed. Reg., Vol. 49, pp. 47714-47770 (December 6, 1984), Washington, D.C.

7. NRC. 1983. "U.S. Nuclear Regulatory Commission, 10 CFR Part 60, Disposal of High-Level Radioactive Wastes in Geologic Repositories--Technical Criteria." Fed. Reg., Vol. 48, pp. 28194-28229 (June 20, 1983), Washington, D.C.

8. E. Gause and P. Soo. 1985. "Review of DOE Waste Package Program, Subtask 1.1 - National Waste Package Program." Vol. 7 of 7 volumes, NUREG/CR-2482; BNL-NUREG-51494, Brookhaven National Laboratory, Upton, New York.

9. A. B. Johnson, Jr., E. R. Gilbert and R. J. Guenther. 1983. "Behavior of Spent Nuclear Fuel and Storage System Components in Dry Interim Storage." PNL-4189, Rev. 1, Pacific Northwest Laboratory, Richland, Washington.

10. R. J. Guenther et al. 1984. "Assessment of Degradation Concerns for Spent Fuel, High-Level Wastes, and Transuranic Wastes in Monitored Retrievable Storage." PNL-4879, Pacific Northwest Laboratory, Richland, Washington.

11. C. N. Wilson. 1985. "Results from NNWSI Series 1 Spent Fuel Leach Tests." HEDL-TME 84-30, Westinghouse Hanford Company, Richland, Washington.

12. R. S. Forsyth, K. Svanberg and L. Werme. 1984. "The Corrosion of Spent UO_2 Fuel in Synthetic Groundwater." In Scientific Basis for Nuclear Waste Management VII, ed. G. L. McVay, pp. 179-190. Mat. Res. Soc. Symp. Ser. Vol. 16, North-Holland Press, New York, New York.

13. L. H. Johnson, N. C. Garisto and S. Stroes-Gascoyne. 1985. "Used-Fuel Dissolution Studies in Canada." In Waste Management '85, ed. R. G. Post, Vol. 1, pp. 479-481, The University of Arizona, Tucson, Arizona.

14. DOE. 1980. "Final Environmental Impact Statement, Management of Commercially Generated Radioactive Waste." DOE/EIS-0046F, U.S. Department of Energy, Washington, D.C.

15. R. A. Palmer et al. 1982. "Characterization of Reference Materials for the Barrier Materials Test Program." RHO-BW-ST-27P, Rockwell Hanford Operations, Richland, Washington.

16. R. P. Anantatmula, C. H. Delegard and R. L. Fish. 1984. "Corrosion Behavior of Low-Carbon Steels in Grande Ronde Basalt Groundwater in the Presence of Basalt-Bentonite Packing." In Scientific Basis for Nuclear Waste Management VII, ed. G. L. McVay, pp. 113-120. Mat. Res. Soc. Symp. Ser. Vol. 26, North-Holland Press, New York, New York.

17. K. M. Harmon, L. T. Lakey and I. W. Leigh. 1984. "Summary of National and International Fuel Cycle and Radioactive Waste Management Programs: 1984." PNL-5138, Pacific Northwest Laboratories, Richland, Washington.

18. D. C. Kocher et al. 1984. "A Perspective on Demonstrations of Compliance for High-Level Waste Disposal." In Waste Management '84, Vol. 1, ed. R. G. Post, pp. 113-117, The University of Arizona, Tucson, Arizona.

19. Office of Nuclear Waste Isolation. 1984. "Performance Assessment Plans and Methods for the Salt Repository Project." BMI/ONWI-545, Battelle Memorial Institute, Columbus, Ohio.

20. Basalt Waste Isolation Project. 1984. "Performance Assessment Plan." SD-BWI-PAP-001, Rockwell Hanford Operations, Richland, Washington.

21. D. H. Alexander et al. 1985. "Conceptual Model for Deriving the Repository Source Term." In Scientific Basis for Nuclear Waste Management VIII, eds., C. Jantzen, J. Stone and R. Ewing, pp. 439-450. Mat. Res. Soc. Symp. Proc., Vol. 44, Materials Research Society, Philadelphia, Pennsylvania.

22. International Atomic Energy Agency. 1983. "Concepts and Examples of Safety Analyses for Radioactive Waste Repositories in Continental Geologic Formations." Safety Series No. 58, Vienna, Austria.

23. Organization for Economic Co-Operation and Development. 1984. "Long Term Radiation Protection Objectives for Radioactive Waste Disposal." Nuclear Energy Agency, Paris, France.

24. R. Rometsch and H. Issler. 1984. "Establishing Repositories for Radioactive Wastes in Switzerland." In Radioactive Waste Management, Vol. 3, pp. 191-203, International Atomic Energy Agency, Vienna, Austria.

25. K. Andersson, N. A. Kjellbert and B. Forsberg. 1984. "Radionuclide Migration Calculations with Respect to the KBS-3-Concept." Technical Report SKI84:1, Swedish Nuclear Power Inspectorate, Stockholm, Sweden.

26. Y. Sousselier. 1983. "National and Cooperative Program for Waste Management in France." In The Treatment and Handling of Radioactive Wastes, eds., A. G. Blasewitz, J. M. Davis, and M. R. Smith, pp. 22-26. Battelle Press, Richland, Washington.

27. K. Uematsu. 1983. "Status of High-Level and Alpha-Bearing Waste Management in PNC." In The Treatment and Handling of Radioactive Wastes, eds. A. G. Blasewitz, J. M. Davis and M. R. Smith, pp. 27-33, Battelle Press, Richland, Washington.

28. R. Papp et al. 1984. "The Technical Concepts for the FRG Alternative Fuel Cycle Evaluation." In Waste Management '84, Vol. 1, ed. R. G. Post, pp. 69-76. University of Arizona, Tucson, Arizona.

29. K. Nuttall, D. J. Cameron and F. P. Sargent. 1982. "The Canadian Engineered Barriers Program." In Proceedings of the Canadian Nuclear Society International Conference on Radioactive Waste Management, pp. 20-23. Winnepeg, Manitoba, Canada.

30. F. S. Feates and H. J. Richards. 1983. "United Kingdom Regulatory Procedures for Radioactive Wastes." In Scientific Basis for Nuclear Waste Management VI, ed. D. G. Brookins, pp. 399-405. Mat. Res. Soc. Symp. Proc. Vol. 15, North-Holland, New York, New York.

31. L. B. Nilsson and T. Papp. 1984. "A Concept for Safe Final Disposal of Spent Nuclear Fuel." In Radioactive Waste Management, Vol. 3, pp. 93-106. International Atomic Energy Agency, Vienna, Austria.

32. Y. Sousselier and F. Van Kote. 1984. "Criteres D'Acceptation Pour de Stockage Definitif Souterrain Des Dechets Radioactifs." In Radioactive Waste Management, Vol. 3, pp. 3-28, International Atomic Energy Agency, Vienna, Austria.

33. K. W. Dormuth and J. S. Scott. 1984. "Research and Development for a Plutonic Rock Radioactive Waste Disposal Vault." In Radioactive Waste Management, Vol. 3, pp. 221-242. International Atomic Energy Agency, Vienna, Austria.

34. L. B. Nilsson and C. Thegerström. 1984. "The Swedish KBS-3 Concept and Program for Continued R&D Work." In Waste Management '84, ed. R. G. Post, pp. 77-84. University of Arizona Press, Tucson, Arizona.

35. J. Chanteur. "Principes Regissant Les Authorisations De Rejet." In Disposal of Radioactive Waste, pp. 171-181. Organization for Economic Co-Operation and Development, Nuclear Energy Agency, Paris, France.

36. A. W. Kenny. "Administrative and Legal Control of the Release of Wastes, Including Monitoring." In Disposal of Radioactive Waste, pp. 183-189. Organization for Economic Co-Operation and Development, Nuclear Energy Agency, Paris, France.

37. P. Dejonghe et al. 1983. "General Perspectives in Radioactive Waste Management in Belgium." In The Treatment and Handling of Radioactive Wastes, eds. A. G. Blasewitz, J. M. Davis and M. R. Smith, pp. 7-13. Battelle Press, Richland, Washington.

SESSION III

LINKAGES BETWEEN MODELS IN SYSTEM PERFORMANCE ASSESSMENTS

THE LINK BETWEEN DETAILED PROCESS MODELS AND SIMPLIED MODELS

K.W. Dormuth and R.B. Lyon
Atomic Energy of Canada Limited
Whiteshell Nuclear Research Establishment
Pinawa, Manitoba ROE 1LO, Canada

ABSTRACT

This paper presents a procedure by which observations of field processes may be translated into mathematical descriptions in the form of Research Interface Models (RIMs) and how these may be used as the basis for the development of Assessment Interface Models (AIMs) for use in overall assessments of the performance of radioactive waste disposal systems. Case examples are used to show how a RIM may be calibrated and validated using experimental results and used to develop an AIM that satisfies the requirements of an assessment model and produces results compatible with those of the RIM over the range of interest.

RELATION ENTRE LES MODELES DETAILLES DE PROCESSUS ET LES MODELES SIMPLIFIES

RESUME

Les auteurs exposent une méthode permettant de convertir les observations des processus in situ en expressions mathématiques sous la forme de modèles d'interface pour la recherche (RIM) et la façon dont ces modèles peuvent servir de base à l'élaboration de modèles d'interface pour l'évaluation (AIM) destinés à être utilisés dans l'évaluation globale des performances des systèmes d'évacuation de déchets radioactifs. Ils font appel à des exemples précis pour montrer comment l'on peut étalonner et valider un RIM au moyen de résultats expérimentaux et l'utiliser pour élaborer un AIM qui remplisse les conditions requises d'un modèle d'évaluation et fournisse des résultats compatibles avec ceux du RIM dans la fourchette de valeurs présentant de l'intérêt.

1. INTRODUCTION

The assessment of the long-term safety of a disposal system for high-level radioactive wastes involves the use of mathematical models to quantify the release and rate of release of contaminants from the engineered and natural components of the system. We will not discuss alternatives to the use of mathematical models, except to say that, to the best of our knowledge, all members of the international community involved in safety assessments are proposing to use them.

The models used fall generally into four categories:

(1) Models to determine the release of contaminants from an engineered waste package. These must deal with processes such as corrosion of metal containers, leaching of fuel or glassified waste, and diffusion through porous buffer materials.

(2) Models to determine the movement of contaminants through the subsurface geological environment. These must deal with processes such as movement of groundwater through rocks and soils, convection and diffusion of contaminants, and chemical reactions between contaminants and geological materials.

(3) Models to determine the movement of contaminants through the near-surface and surface environments. These deal with deposition and transport in surface waters and sediments and uptake by plants and animals.

(4) Models to determine the dose to biological species resulting from exposure to, or ingestion of, the contaminants. In many cases these models deal specifically with man as an indicator of the dose to other biological species.

Much of the research in support of radioactive waste disposal has the objective of developing these models on a firm scientific foundation.

2. TYPES OF MODELS

If the results of research programs are to be incorporated in the safety assessment through the use of mathematical models, the models must have the following characteristics:

(1) The models must interface well with the experiments and the experimenters. Therefore, they must represent the processes being studied in sufficient detail and with sufficient accuracy that they can be used to interpret the research results through comparison of calculated results with observations, and their input must be determined straightforwardly from measurements.

(2) The models must interface with other models used in the assessment and be consistent with them. This requires compatible sets of input and output parameters, or, equivalently, interface routines to link the models.

(3) The models should be constructed so that important elements of the input (geometry, boundary conditions, material properties, etc.) can be easily changed. Also, to be practical, they should be reasonably economical in terms of computer resources. This is essential because they will be used to perform sensitivity analyses and/or uncertainty analyses to determine the variability in the performance of the disposal system.

(4) The models should be understandable to as large as possible a cross-section of the scientific community, who must be satisfied that the results of the calculations can be used as a basis for decisions on the acceptability of waste disposal facilities and concepts.

Possibly a single model of a physical process could satisfy the requirements of both a model to interface with experimental programs (requirement 1) and a model to be used in the safety assessments (requirements 2-4). However, it is unlikely that a single model could meet all requirements efficiently or effectively. In general, we consider that two types of model must be involved in connecting field and laboratory research to the safety assessment. The first type is developed with relatively little thought to satisfying requirements 2, 3 and 4, but is developed primarily to interface effectively with field and laboratory experiments. We call this the Research Interface Model (RIM). The second type is developed primarily to act effectively as a submodel in an overall safety assessment, with particular attention to requirements 2, 3 and 4. We call this the Assessment Interface Model (AIM).

In general, the procedure for producing an assessment is to develop and validate a RIM to analyse a range of problems to be considered in the assessment. The results of these analyses are used to develop an AIM, which produces results compatible with those of the RIM over the sets of conditions required for the assessment. This ensures that extrapolations beyond actual experience are made with a model (AIM) that has been shown to give similar results for representative sets of conditions as a model (RIM) that has been shown capable of valid predictions by comparison with experiments. The AIM, which in general cannot be compared directly with experimental observation, is used only to interpolate within the range for which it has been compared with the RIM. This validation procedure is concerned only with the analysis of the particular subsystem to which the AIM is applied. It does not validate the interfaces between the AIM and models, or the other subsystems. Discussion of this latter aspect of validation is beyond the scope of this paper.

In the following section, we give some examples of the development of AIMs for use in various assessments.

3. CASE EXAMPLES

3.1 Dissolution of Used Fuel

Goodwin et al. (1982) developed an AIM to model the dissolution of used UO_2 fuel in the expected environment of a disposal vault in plutonic rock. It was assumed that the process consisted of an instantaneous release followed by a slower release congruent with the dissolution of the UO_2 matrix. The justification for these assumptions is discussed in the reference.

One of the parameters required was the solubility of UO_2 under disposal vault conditions. Paquette and Lemire (1981) used a modified version of a computer program developed by Froning and Verink (1976) to construct uranium solubility isopleths over the range of Eh and pH values expected in the vault environment, based on fundamental thermodynamic data. Information on the distribution of Eh and pH values encountered in plutonic rock environments was then used to obtain a curve of cumulative probability versus uranium solubility in a disposal vault at 100°C. The uranium solubility in the AIM was obtained by sampling from this distribution.

The RIM in this case is a detailed thermodynamic model used to develop the potential-pH diagrams. The validity of this model to derive solubilities under the stated assumptions is a matter for establishment by the specialists. The AIM itself cannot be compared directly with experiment, but is valid as long as the RIM is, since it merely selects from the results derived from the RIM.

An alternative approach, developed to more rigorously take into account the effect of the geochemical environment on all the release and sorption processes in a disposal vault, to expresses the rate of these processes as a function of the geochemical parameters controlling them (Lyon and Garisto, 1984). For the solubility-controlled dissolution of used fuel, a simple mathematical formula (i.e., an AIM) has been derived, for the calculation of UO_2 solubilities as a function of temperature, pH, oxidation potential, and ligand concentrations (Garisto and Garisto, 1985a). The solubility of UO_2 in a disposal vault can, therefore, be determined without the need for complex thermodynamic equilibrium programs (i.e., a RIM). A similar AIM has been developed for U_4O_9 solubilities, taking experimental results of radiolysis into account (Garisto, 1985; Johnson et al., 1985). The AIMs for UO_2 and U_4O_9 solubilities are based on an updated thermodynamic database for uranium (Lemire, 1985) developed by the OECD Nuclear Energy Agency (Muller, 1985).

The accuracy of the UO_2 solubility formula (an AIM) was established through a comparison with full thermodynamic calculations (RIMs) by Garisto and Garisto (1985a). In addition, the oxidative dissolution of UO_2 has been simulated using thermodynamic reaction path models (other RIMs). The theoretical predictions of these RIMs were compared with available results of electrochemical dissolution experiments (Garisto and Garisto, 1985b). The agreement between the predictions of the RIMs and UO_2 dissolution experiments supports the use of thermodynamically derived source terms in environmental and safety assessments of used fuel disposal.

3.2 Transport Through The Geosphere

As part of the International Fuel Cycle Evaluation, an analysis was done of the potential release from a hypothetical geologic radioactive waste vault in hard rock (INFCE, 1979). The AIM in this case was a variation of the GETOUT model (Lester et al., 1975), which was used to calculate the migration of radionuclides through the geosphere. It is a one-dimensional transport model that requires lumped parameters, such as average groundwater transit and path lengths, to describe convection, dispersion, and geochemical retardation.

The actual system to which the assessment is to apply is fractured rock. The water moves through a complex network of fractures and networks of channels within those fractures. It is driven primarily by head gradients caused by variable topography and heat from the vault.

To relate the complex behaviour of the actual system to the simplified picture in the AIM, a finite-element, three-dimensional, groundwater-flow computer code, FE3DGW (a RIM) was applied to some data taken from an actual field site. The model assumed that flow through the fractured rock could be described by treating the rock as a heterogeneous porous medium. Variability of hydrogeological properties with depth was taken into account by assigning different values of permeability and porosity to elements at different depths. Large-scale subvertical linear features (faults) observed at the field site were treated by modelling them with narrow elements that could be given permeabilities and porosities different from the surrounding rock. The driving force generated by the topographic gradient was produced by using head distributions on the upper boundary, derived from a topographic map of the field area. Hydrogeological parameters in the model were varied to determine sensitivity of the model to the parameters.

The average groundwater transit times from the hypothetical vault to the surface and the average path lengths calculated by the RIM, FE3DGW, were used in the AIM. The validity of the RIM was not fully established in this case by comparison with observations at the actual field site.

In this example, a RIM was used to provide input values to a preconceived AIM. However, we believe that, in general, it should not be assumed that a particular AIM is capable of assimilating information from particular field or laboratory observations into an assessment. A more general procedure for progressing from experimental observation to an assessment model is to develop and validate a RIM based on observation, and then to conceptualize an AIM that calculates the important aspects of the physical system in a way that gives results in agreement with those of the RIM over the required range.

The direct interaction of a RIM with a field research program is illustrated by the drawdown experiment conducted at the site of the Underground Research Laboratory (URL) near the town of Lac du Bonnet, Manitoba, Canada. The experiment is a large-scale validation exercise that tests the ability to use information obtained from the surface and from boreholes to model the hydrogeological behaviour of a plutonic rock mass on the kilometre scale.

The experiment (Davison and Guvanasen, 1985) consists of four phases: (1) geological and hydrogeological characterization of the URL site before excavation of the shaft, (2) development and calibration of hydrogeological models of the site based on pre-excavation observations, (3) prediction of the changes in the groundwater regime due to excavation of the URL, and (4) comparison of predictions with observations of the changes in the groundwater conditions during and subsequent to excavation. The first three phases are complete, and preliminary results are available from the fourth phase.

Atomic Energy of Canada Limited applied the three-dimensional, finite-element, hydrogeological computer program, MOTIF, to the analysis. (Other groups participated by performing analyses with other codes.) The model was calibrated using long-term head measurements and the results of inter-borehole pumping tests, and the calibrated model was used to predict the changes in the groundwater regime that would occur due to excavation of the URL shaft.

Preliminary comparisons of model predictions with observed changes in the hydrogeological conditions have been completed. The change in piezometric conditions in the rock mass with time was, in general, predicted well by the model. Discrepancies did occur very near the excavation where the model under-predicted the drawdown. However, at most observation points, the predicted spatial and temporal distributions of the drawdown were in good agreement with observation. The overall trend in the predicted rate of inflow into the excavation also agreed well with observation. The magnitude of the inflow was, in general, over-predicted by a factor of three. The discrepancies are attributed to several factors, including deviations of actual excavation rates from assumed rates, the existence of vertical fractures connecting the shaft with a highly conductive subhorizontal fracture zone, and the effect of stresses on the local hydraulic conductivity of the rock caused by excavation. It should be emphasized that the discrepancies are very small compared with the variability in values of hydraulic parameters in the rock mass. The overall agreement between observation and prediction is considered to be extremely good.

The experiment clearly establishes the relationship between the MOTIF RIM and information derived from field testing and measurement. It also gives evidence to support the ability of the RIM, as employed by field researchers and modellers, to predict the behaviour of the hydrogeological system on a large scale. Validation of MOTIF is being further developed by comparing the code with analytical solutions of appropriate idealized systems and with other computer models.

MOTIF is currently being used to model a plutonic rock body containing a hypothetical disposal vault. The rock structure and hydrogeological conditions are conceptualized using data from the research area surrounding the URL. Using the results of the MOTIF analysis, it is intended to generate an AIM to model transport through the geosphere, as part of an overall disposal system assessment. The structure of the AIM is not pre-determined but will be developed to be compatible the conceptual model of the particular area and with the results generated by the RIM.

4. CONCLUSION

A general procedure has been described for developing a model based on the results of field and laboratory research to be used as part of an assessment of a disposal system. A RIM is calibrated and validated using experimental results, and used to develop an AIM that satisfies the basic requirements of an assessment model and produces results compatible with those of the RIM over the range of interest.

5. REFERENCES

Goodwin, B.W., R.J. Lemire and L.H. Johnson (1982), "A Stochastic Model For The Dissolution Of Irradiated UO_2 Fuel," Proceedings of the CNS International

Conference On Radioactive Waste Management, Winnipeg, Manitoba, Canada, 1982 September 12-15, pp. 298-304.

Davison, C.C. and V. Guvanasen (1985). "Hydrogeological Characterization, Modelling and Monitoring Of The Site Of Canada's Underground Research Laboratory," Proceedings of the 17th International Congress of the International Association of Hydologist on Hydrogeology of Rocks of Low Permeability, Tucson, Arizona, 1985 January 7-9.

Froning, M.H. and E.D. Verink, Jr. (1976), "A Computer Programme for Calculation of Potential-pH Diagrams of Metal-Ion-Water Systems," NTIS, AD-A027634, University of Florida at Gainesville.

Garisto, F. and N.C. Garisto (1985a), "A UO_2 Solubility Function for the Assessment of Used Nuclear Fuel Disposal," Nucl. Sci. Eng. 90, pp. 103-110.

Garisto, N.C. and F. Garisto (1985b), "The Dissolution of UO_2: A Thermodynamic Approach," Atomic Energy of Canada Report AECL-8887, and references therein. Submitted to Nucl. Chem. Waste Management.

Garisto, N.C. (1985), "Modelling of Used Fuel Dissolution," In Proceedings of the Nineteenth Information Meeting of the Nuclear Fuel Waste Management Program, 1985 May, Toronto, Atomic Energy of Canada Technical Record Report, TR-350* in press.

INFCE (1979). "Release Consequence Analysis For A Hypothetical Radioactive Waste Repository In Hard Rock," INFCE/DEP/WG7/21, IAEA, Vienna.

Johnson, L.H., N.C. Garisto and S. Stroes-Gascoyne (1985), "Used-Fuel Dissolution Studies In Canada," In Proceedings of Waste Management '85 Tucson, Arizona, 1985 May 24-28.

Lemire, R.J. (1985), Private communication.

Lester, D.H., G. Jansen and H.C. Burkholder (1975), "Migration Of Radionuclide Chains Through An Adsorbing Medium," Amer. Inst. Chem. Eng. Symposium Series 71, 202-213.

Lyon, R.B. and N.C. Garisto (1984), "Environmental Modelling and Geologic Disposal of Nuclear Fuel Waste," In Mineralogical Association of Canada Handbook, Vol. 10, Ed. M.E. Fleet.

Muller, A.B. (1985), "International Chemical Thermodynamic Data Base for Nuclear Applications," In Radioactive Waste Management and the Nuclear Fuel Cycle 6 (2), pp. 131-141.

Paquette, J. and R.J. Lemire (1981), "A Description Of The Chemistry Of Acqueous Solutions Of Uranium and Plutonium To 200°C Using Potential-pH Diagrams," Nucl. Sci. Eng. 79, 26-48.

* Unrestricted, unpublished report available from SDDO, Atomic Energy of Canada Limited Research Company, Chalk River, Ontario KOJ 1J0.

THE LINKS BETWEEN INDIVIDUAL PROCESS MODEL AND INTEGRATED MODEL IN THE PROBABILISTIC RISK ASSESSMENT CODES

A. Saltelli
Commission of the European Communities
Joint Research Centre - Ispra Establishment

ABSTRACT

One of the most complex issues related to the implementation of a system performance analysis code is given by the links among the submodel in the integrated system. The problem is discussed here with reference to the architecture of the existing codes and to their boundary conditions. The role of correlation in ensuring system consistency is also touched upon.

RELATIONS ENTRE LES MODELES SIMULANT DES PROCESSUS INDIVIDUELS ET LE MODELE INTEGRE DANS LES PROGRAMMES DE CALCUL APPLICABLES A L'EVALUATION PROBABILISTE DES RISQUES

RESUME

Les relations entre les sous-modèles utilisés dans le système intégré constituent l'une des questions les plus complexes que pose la mise en oeuvre d'un programme de calcul relatif à l'analyse des performances des systèmes. Le problème est analysé du point de vue de l'architecture des programmes de calcul existants et de leurs conditions aux limites. Le rôle de la corrélation dans le maintien de la cohérence du système est également évoqué.

1. INTRODUCTION

The assessment of nuclear waste disposal impact on the environment involves complex and integrated models to evaluate the possible consequences of radioactivity release events. Computer codes have been developed which incorporate such models in a Monte Carlo scheme to estimate both the release consequence and the consequence likelihood accounting for the parameter uncertainty and variability. Three such computer programs are documented in the literature:

- NUTRAN, developed at Lawrence Livermore Laboratory, early in 1979 (Ross, 1979);
- SYVAC, of the Atomic Energy of Canada Limited, first developed in 1981 (Dortmuth, 1981);
- LISA, recently developed at the Joint Research Centre of the European Communities (Saltelli, 1984, 1985).

Considerable work in the application of a Monte Carlo methodology to waste disposal systems has been done by the group of the SANDIA Laboratories, especially in relation to the statistical framework of the system: sampling techniques, use of response surface methodology, correlation among the input variables and so on (Iman, 1980; Helton, 1981).

Other codes of the same family are likely to become available in the near future as a number of groups in the Waste Management community are developing their own assessment tool. At the same time the number of versions of the already existing codes is increasing, due to the extension of the original models to new sites and storage options, to the model updating and to new experimental evidences.

The topic addressed here is general for this class of codes and concerns the links between the various submodels in the integrated systems. Setting aside - for the moment - all the statistical frame required by the Monte Carlo methodology (e.g. the subroutines for the sampling, the correlation, the sensitivity analysis, etc.), a code for the Performance Assessment (PA) can be considered as a chain of mass transfer submodels. Each submodel is meant to simulate the transfer of activity through a "barrier", i.e. through a part of the complex system which separates the waste in its underground disposal from a man living on the earth surface. Although such a subdivision of the real world into a number of subsystems is very effective from the computing point of view, it is de facto an approximation. Mass transfer, in fact, occurs simultaneously in the various barriers and the model equations describing the system should be solved simultaneously. Breaking down the system into a number of subsystems results, instead, in the subdivision of the original set of equations into a number of subsets, which are linked through a series of approximated boundary conditions. This is not the only "decoupling" process which is practiced in PA modelling. Even within the single subsystem (or barrier) phenomena which are intrinsically associated are often modelled separately for sake of simplicity.

In this note the implications of the model decoupling are discussed to some extent. Also the testing of the overall PA codes will be addressed with reference to the links between the various subsystems. The last section is dedicated to the use of correlation in ensuring system consistency. A system using randomly selected input vectors becomes inconsistent when physically impossible (or highly unlikely) combinations of input parameters are allowed to enter the system. In particular, a pessimistic - but conservative - choice of a parameter value can be exclusive with respect to a parallel choice of another parameter. Parameter correlation can be imposed to account for the degree of likelihood of critical parameter combinations.

2. DECOUPLING OF PROCESS AND ARCHITECTURE OF THE INTEGRATED SYSTEM

In performance assessment codes, complex transfer systems are modelled where, for instance, contaminants are leached out of a matrix, cross a container barrier and enter a buffer material; then they pass to the host formation and to a system of aquifers up to the geo-biosphere interface(s); here they enter the different contamination streams through water bodies, air, soils, vegetables, cattle, etc., up to man.

An "exact model for such a system is hardly conceivable. Not only should such a model contain all the pertaining "exact" equations, but these should be solved simultaneously. To make an example, near-field and far-field should be considered as a "continuum", and in the numerical solution of the equations describing the various submodels, there should be a single space mesh, with steps of different sizes, depending on the accuracy-stability requirements of any single barrier (Fig. 1). In this form the system needs, apart from the initial conditions, only two boundary conditions (input and output), at least as long as the highest space derivative is of the second order.

It may be worth noting that the simultaneous solution of the complete set of equations with an analytical procedure is not practicable. Furthermore, in the

Fig. 1

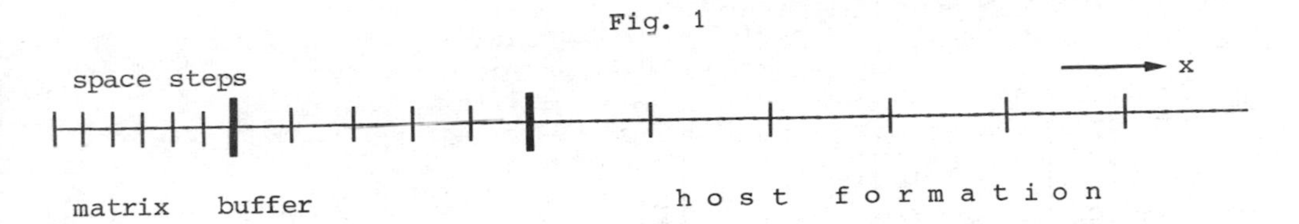

numerical scheme, there will be serious time-stepping problems, as the coarser space mesh - or anyhow the most unfavourable stability condition - will impose the time-step to be used on the entire mesh, both if the integration is being performed with finite differences method (LISA) or with the Green functions method (NUTRAN, SYVAC). Other shortcomings of the above system are a general lack of flexibility and the difficulty of testing. For these reasons it is common practice to break down the system into a number of separate modules, generally one module for each barrier, and to solve the system equations for each module separately, the output from a single barrier providing the input to the next one. This simplification, on the other hand, implies an "inherent" error, which is that the ith module does not know what is happening in the $(i+1)^{th}$ one. More precisely, the $(i+1)^{th}$ module receives its input from the ith one, but the output boundary condition for the ith module does not account for the concentration (or activity or whatever) in the subsequent block. Dividing the system in blocks has resulted in increasing the number of the boundary conditions, but these conditions are inherently wrong. The error in this case is in the links between the submodels and it is conceptually higher when the downstream block is seriously affecting the upstream one. A classical

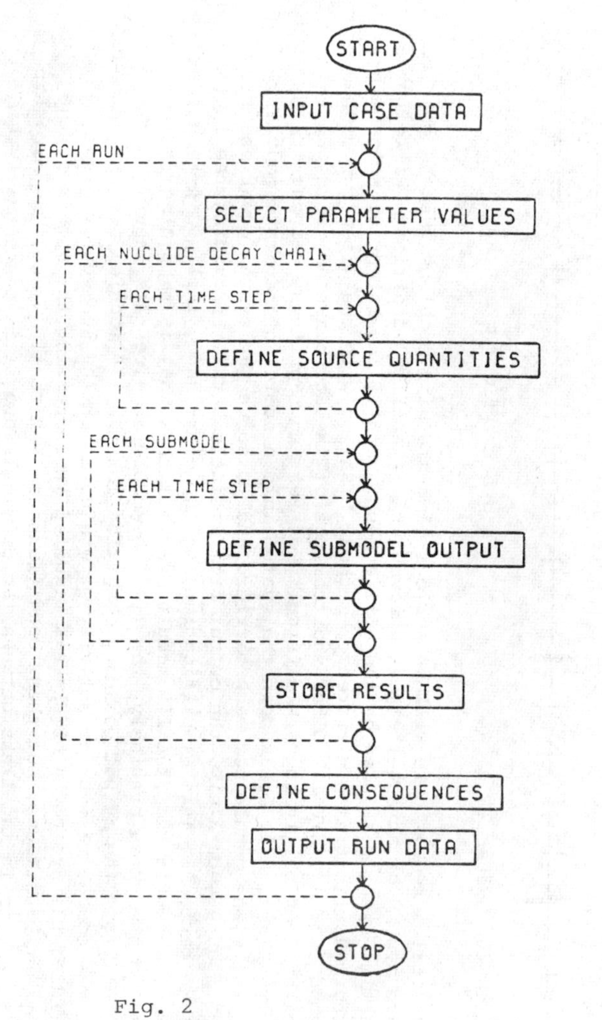

Fig. 2

(from Sherman, 1985)

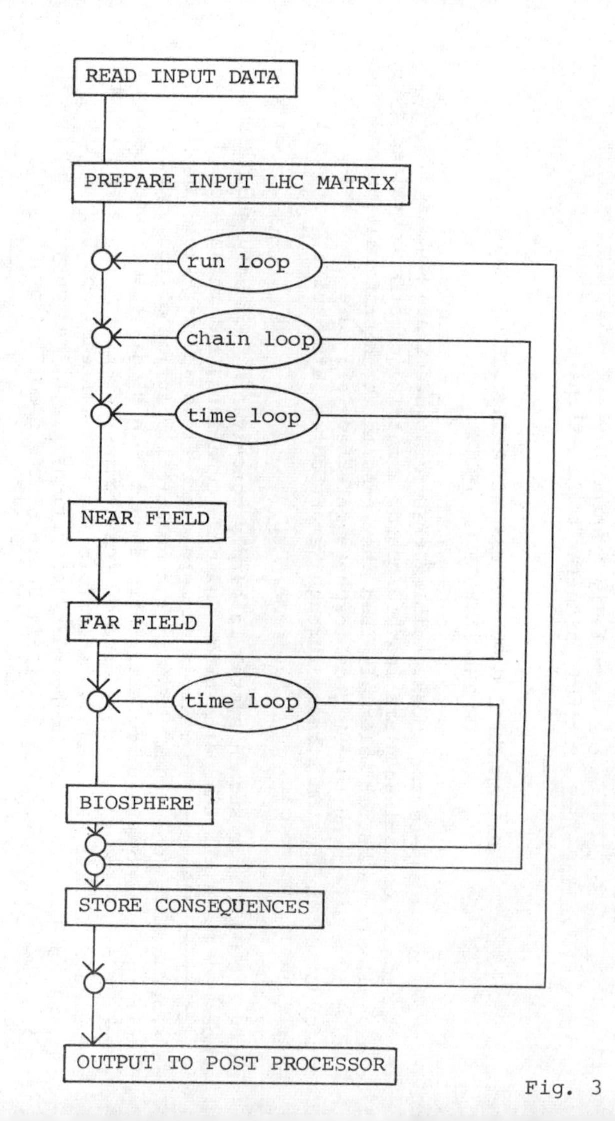

Fig. 3

example might be that of glass in contact with a clay buffer: clay enhances glass matrix dissolution and has a very selective action on the radioelements contained in the glass and on the glass constituents. The decoupling of the system glass-buffer must then be operated very carefully, keeping in mind all the known synergetic effects and introducing them into the system through the appropriate boundary conditions. Conservatism must provide guidance in such a process. A discussion of the above problem is given in the section dedicated to the boundary conditions.

A second class of inherent errors is linked to the single submodels. Within a single barrier there will be, in general, a number of coupled phenomena: heat transfer, mass transfer, chemical reactions, etc. Although any barrier can be modelled, in principle, with the required degree of accuracy, the models to be used in the performance assessment codes must be kept simple, at least as long as Monte Carlo methods are being used. It is not viable, for instance, to introduce in SYVAC or in LISA a three-dimensional hydro-geochemical code or to model heat and mass transfer simultaneously. Even here, decoupling of process results in both a simplification and an inherent error.

An inherent error in the model does not necessarily imply an error in its actual prediction. For certain combinations of the input parameters, the assumptions leading to the simplification may work well. For instance, for a wide range of Darcy velocities in the aquifer, the flow host-rock/aquifer is independent from the nuclide concentration in the aquifer. Unfortunately, the Monte Carlo codes used in PA are meant to run over extremely large ranges of input values, due to the uncertainty affecting both scenarios and parameters, so that errors arising from both phenomena and barriers decoupling are unavoidable.

The only viable strategy in this case is to force the error in a pessimistic direction, choosing the most unfavourable ranges of value for the pertaining parameters. For instance, the leach rate of the waste form can be given a distribution which is pessimistic enough to account for all the expected or suspected synergetic effects. Correlation techniques can then be used to mitigate the pessimism of certain combinations of input parameters, avoiding, for instance, that a high fresh water/soil transfer coefficient is sampled together with a low soil K_D for the same nuclide.

Once the subdivision of the system in subcomponents is made, and the submodels have been elaborated, the architecture of the integrated system has to be chosen. The Monte Carlo codes used in PA involve three main loops:

- run loop: the same calculations are repeated for different combinations of the input parameters;
- nuclide chain loop: the same calculations are repeated for the different nuclide chains;
- time loop: the same calculations are repeated for all the times up to the cut-off time.

The architecture of a system is mainly determined by the arrangement of the various submodels within these loops.

SYVAC and LISA structures are schematized in Figs. 2 and 3, respectively. The advantages of these structures are an easy module replacement, the possibility of storing intermediate outputs and an easier error debugging. Both systems allow the possibility of restarting the simulation if it were halted by an error in a certain run.

3. BOUNDARY CONDITIONS

Individual process models are linked in the integrated system through a set of boundary conditions. The output from a given submodel coincides or is generally included in the input boundary condition for the submodel immediately downstream. At the same time an output boundary condition is generally required for each submodel. An exact boundary condition should take into consideration the concentration of the migrating species in the first cell of the next box. As mentioned before, this is not feasible when the equations are being solved separately for each submodel. This problem is susceptible to different approaches; three examples, taken from LISA, SYVAC and NUTRAN, are discussed here.

LISA

The two interfaces buffer/geosphere and geo/biosphere are taken as an example. The buffer is - in LISA - a purely diffusive barrier, with no water movement. Typically, a layer of 0.5 - 1 m of bentonite. In the present version there is no backfilling, so that after the buffer the contaminants enter directly the host formation, where a certain convective transport is allowed. In this scheme, it is reasonable to assume that the rate of removal of nuclides from the interface is higher than the rate at which they are supplied at the interface from the bulk of the buffer. Consequently, an infinite dilution at the end of the buffer is assumed, i.e. nuclide concentration is taken as zero at the interface. This condition is always conservative, as it maximizes the output flux. A similar condition is taken at the geo/biosphere interface, either a river or a well. In this case, the infinite dilution condition appears to be quite natural. It can be noted that such a condition becomes too pessimistic when saturation effects are expected, but it remains conservative as long as synergetic phenomena are excluded.

SYVAC

In SYVAC-2 the vault buffer material is assumed to allow both diffusive and convective transport. In this way the set of equations describing the transport in the buffer is very similar to the one describing the geosphere, and the transport is governed by the same mechanisms in the two media. A zero concentration boundary condition is not applicable here, as the geosphere permeability and hydraulic gradient might be not high enough to keep the interface clean.*

The solution adopted in SYVAC 2 is to equate the flux at the interface to the interfacial concentration times a mass transfer constant, i.e.

$$- D\frac{\partial c}{\partial x}\bigg|_{x=a} + V\, c(a,t) = K_V\, c(a,t)$$

where:

c = nuclide concentration, M

a = interface coordinate, m

D = dispersion coefficient, m^2/a

K_V = mass transfer constant, m/a.

---

* A zero concentration boundary condition was used in SYVAC-1 (Wuschke, 1981).

The mass transfer coefficient K_V is defined as

$$K_V = \frac{\tau_B q_T}{\varepsilon_B}$$

where:

τ_B = tortuosity factor of the buffer, dimensionless

q_T = flow velocity in the geosphere, m/a

ε_B = buffer porosity, dimensionless,

and accounts for the relative values of the water flow in the two media (LeNeveu, 1985). It can be noted that this formula is, in fact, an approximation, as it does not include the element concentration in the geosphere, but it allows the buffer model to know which flow regime is taking place beyond the interface.

NUTRAN

NUTRAN is the oldest code of the family and - in many respects - the most approximate. Just to make an example, the retention constants for all the members of a decay chain are assumed equal, to allow a simpler solution of the geosphere transfer model with the Green functions method. The transport code WASTE, built in NUTRAN, approximates the ground water flow pattern around a repository by a network of one-dimensional flow paths. This provides a flexible and effective way of computing water flow rates to simulate multidimensional flow paths, accounting for different repository geometries. For the mass transfer a zero concentration boundary condition is implicitly assumed at the end of each elementary path.

4. TESTING OF THE INTEGRATED SYSTEM

A preliminary remark about the testing of the integrated system might be the Murphy Law of Programming which states: There is always one more bug. More constructively, it can be said that only a system of well-tested individual modules has some chance of providing sensible results after a finite time of debugging.

Although the accent is here on the testing of the integrated system, a few preliminary remarks can be made on the testing of the individual models:

i) Each module has to be tested as a "stand alone" before inserting it in the code.

ii) The test must involve the same wide range of input parameters which the module is expected to receive in the code.

iii) If the module is a mass transfer one (e.g. buffer, geosphere, etc.), it must be extensively tested for mass balance.

Especially when submodels have been developed by different modellers, the task of creating a coherent integrated system is in many respects a communication problem: communication between modellists, to achieve a mutual understanding of the features and the requirements of each individual module and communication between the modules themselves, to ensure that the proper information is being transmitted throughout the submodels. The classical tool for solving this class of problems is given by the software engineering techniques, where the software quality assurance is achieved through a formalization of the implementation procedure. Quality assurance techniques have been applied in SYVAC to optimize the quality of the software through a formalized implementation procedure. A software quality assurance technique can be defined as "a systematic pattern of activities to establish confidence that the software meets specifications and software performs acceptably" (Andres, 1984). Although a project to rewrite SYVAC using quality assurance techniques was set up early in 1982, time scheduling problems resulted in a redefinition of the project. On the one hand the present SYVAC-2 was developed for the concept assessment without the quality assurance. On the other hand a quality assured code to analyse sensitivity (ANSENS) was produced. According to Andres the most important datum gained through the SYVAC and ANSENS experience is that these software engineering techniques have to be tailored to the working environment of the performance assessment codes. This is mainly due to the rapidity with which such codes evolve as a result of model updating and experimental new advances, so that a detailed specification of the problem - generally required in quality assurance - is not always possible. Although expensive and highly time-consuming,software engineering techniques have been proved to perform effectively. Quality assurance of the software is being foreseen for the future SYVAC-3.

Code intercomparison is an effective tool for checking the correctness of the submodels: in this respect, Intracoin and Hydrocoin have been, and are, respectively, of crucial importance in model testing and development. As far as the integrated systems are concerned, an intercomparison at the level of integrated systems - SYVAC against LISA - is presently under way. The main rationale for this exercise is to point out which differences in the system predictions can be expected by using different models with the same input data. By its own nature this intercomparison is more a test of the philosophy behind the codes than a check of the accuracy of the codes themselves.

On the other hand, this kind of exercise is likely to reinforce the confidence in the model predictions, as they allow a better understanding of the overall performance of the system and point out which are the critical assumptions made in the modellization. It should also be said that integrated systems intercomparison is much more committing than the individual model one. Not only does a common data base have to be agreed upon, but the submodels specifications must be carefully defined to ensure that the two systems are simulating the same kind of phenomena. This involves a good deal of systems readjustment.

Apart from the general software quality assurance techniques there are a number of tests which can be easily applied to the integrated system. Here is a list of simple operations which were found to be useful in the debugging of LISA:

i) Dimensionality check; starting from the moles of a given element per unit of waste form check that the proper units for the output consequence (e.g. sV/a) are obtained.

ii) Test of the system for limiting values of the input parameters, such as for instance: set the inventory of a parent nuclide to zero and check that the element does not show up in the output. Set the decay constant of an element to zero and check that its daughters do not appear in the output. Analogously solubilities, consumption rates, transfer factors, can be set to zero to make sure that the system is working properly.

iii) Back of the envelope calculation; the dose resulting from one or more significant runs can be checked - as order of magnitude - by roughly estimating transit times through the different barriers and the attenuation of the pulse. In doing so it can be useful to neglect dispersivity - where possible - and to assume the steady state for the transfer between the various compartments.

The tests described above are rather trivial. Nevertheless, experience shows that very often - when used - they point out unforeseen errors.

An error can also originate when the code is operative. In this case it is important that the error is detected, possibly during the execution itself, so that the proper corrective action can be taken. Control statements can be introduced to check that a variable is not sampled outside its range or that a concentration does not become negative. Other control statements may involve the stability conditions, the accuracy of an interpolation, that of an integration, or even the correctness of the solution of an eigenvalue problem.

It should be said, anyhow, that the establishment of a level of confidence in the system predictions will never be a completely automatic or formalized process, not even at the level of the system accuracy. In the very end, confidence relies on the knowledge of the system which is acquired by users and modellists through the implementation, the testing and the operation of the code itself.

5. CONSISTENCY OF THE INTEGRATED SYSTEM AND THE ROLE OF CORRELATION

Both in SYVAC and in LISA, the assessment of the contaminant flow impact is done stochastically. Each simulation is composed of a number of runs (or scenarios in the SYVAC terminology). In turn, each run is uniquely determined by the set of sampled values of its input, i.e. by the input vector. Because of the large variability of the input data, a number of such vectors may contain combinations of parameter values which are physically inconsistent. These can also result from the systematic adoption of conservative distributions (and assumption) done in generating the integrated system. As a result, these impossible - or unlikely - input vectors are often biased on the pessimistic side, so that unrealistically high values of the system consequences are predicted. A corrective action must be taken to prevent this type of occurrence in the sampling.

A natural strategy of tackling the problem is to correlate the input variables. In particular, a number of variables might be naturally correlated by making them dependent upon a set of "basic parameters". Such an approach is presently being considered for the SYVAC-3 vault submodel, where, for instance, pH, electrochemical potential, temperature and ionic strength could be taken as basic parameters (Goodwin, 1983). It is assumed that "these parameters contain the uncertainties associated with corrosion, dissolution, etc., then these parameters, together with suitably defined probability distributions, become SYVAC's randomly selected parameters. The second step in this approach requires the provision of functional expressions which relate, for example, penetration time (container lifetime) to the basic set of parameters. The net effect would be to select parameters for the vault which display the required degrees of correlation. Similar approaches might also be useful for the geosphere and the biosphere submodels, possibly using different sets of basic parameters and probability distributions." (Goodwin, 1983).

Another way of correlating variables is by explicitly imposing the desired correlation coefficient between two or more variables. In SYVAC-2 a parameter with normal or log-normal distribution can be defined as correlating with another parameter having one of these distribution types (Sherman, 1985). In this procedure, only the independent parameter is sampled, and the dependent one is computed. This procedure is likely to be somehow too rigid to be extensively employed. As an alternative, joint distributions might be used for the variables to be correlated, but such an approach has not yet been taken in performance assessment codes.

Another possibility is given by a technique developed by Iman and Conover (Iman, 1982), and presently implemented in LISA. With this technique any combination of parameters can be correlated in LISA at the desired level (expressed by a rank correlation coefficient), regardless of the type of individual distribution. The characteristics of this technique are the following:

i) All the correlated parameters are sampled.
ii) The ranks of the sampled values conform with the requested ones.
iii) The individual (marginal) distribution of the variables is not changed by the correlation.

The advantage of this method is that it is quite general and easy to use. It can also be used to purge spurious correlations between the input variables, when this is desired. A criticism which is addressed to the use of this technique for eliminating unwanted correlations is that spurious correlations do occur in nature.

In conclusion it can be said that the integrated system can be made more consistent by correlating the input variables. Concerning the various strategies:

i) The basic variable approach foreseen for SYVAC-3 is very demanding in terms of modelling the relationships between the dependent parameters and the basic set.
ii) The method presently implemented in SYVAC-2 is not flexible enough.
iii) The system implemented in LISA is both general and flexible, and its use should be recommended.

This matter will also be discussed in a number of Specialists Meetings, both within the PAGIS action and the NEA coordinated System Variability Analysis Code Users' Group.

ACKNOWLEDGEMENTS

The Author wishes to thank T. Andres of Atomic Energy of Canada Ltd. for his help in revising this note.

REFERENCES

Andres, T.H., Reid, J.A.K. and Hoffman, K.J., "Software quality assurance in environmental assessment", in: Proc. of the 16th Information Meeting of the Nuclear Fuel Waste Management Program, AECL report TR 218 (1984).

Dormuth, K.W. and Sherman, G.R., "SYVAC - a computer program for assessment of nuclear fuel waste management systems, incorporating parameter variability", AECL report 6814 (1981).

Goodwin, B.W. et al., "SYVAC development", in: Proc. of the 16th Information Meeting of the Nuclear Fuel Waste Management Program, AECL report TR 218 (1984).

Iman, L.R. and Conover, W.J., "A distribution-free approach to inducing rank correlation among input variables", Commun. Statist.-Simula. Computa., 11(3) (1982).

LeNeven, D.M., "Vault submodel for the second interim assessment of the Canadian concept for nuclear fuel waste disposal; post closure phase", Atomic Energy of Canada Ltd. report (in preparation, 1985).

Saltelli, A., Bertozzi, G. and Stanners, D.A., "LISA - a code for safety assessment in nuclear waste disposal", Commission of the European Communities, EUR report 9306 EN (1984).

Saltelli, A., "Improvements and extensions of the LISA code", Commission of the European Communities, JRC technical note 1.06.C2.85.56 (1985).

Sherman, G.R., Donahue, D.C., King, S.G. and So, A., "SYVAC-2: a system variability analysis code for assessment of nuclear fuel waste disposal concepts", Atomic Energy of Canada Limited, report TR 317 (1985).

Wuschke, D.M. et al., "Environmental and safety assessment studies for nuclear fuel waste management", Vol.3, Post-Closure Assessment, AECL report TR 127 3 (1981).

Ross, B. et al., "NUTRAN: a computer model of long-term hazard from waste repositories", UCRL 15150 (1979).

Iman, R.L., Conover, W.J. and Campbell, J.E., "Risk methodology for geological disposal of radioactive waste: small sample sensitivity analysis", SAND80-0020; NUREG/CR 1397 (1980).

Helton, J.C., Brown, J.B. and Iman, R.L., "Risk methodology for geological disposal of radioactive waste: asymptotic properties of the environmental transport model", SAND79-1908; NUREG/CR-1636 (1981).

THE INTEGRATION IN A GLOBAL MODEL OF DETAILED MODELS FOR THE WASTE REPOSITORIES DETERMINISTIC RISK ASSESSMENTS

J. Lewi
Commissariat à l'Energie Atomique
Institut de Protection et de Sûreté Nucléaire
Centre d'Etudes Nucléaires de Fontenay-aux-Roses
92265 Fontenay-aux-Roses Cedex

Abstract

Two methodologies are usually considered when performing assessment calculations of waste repositories : probabilistic and deterministic methods. The "probabilistic" methodology involves the integration of simplified models while the "deterministic" methodology is able to integrate detailed models.

The French approach utilises some of the characteristics of both methodologies. In this report, we describe, on the basis of the MELODIE code development experience, some problems which appear in the integration of detailed models in the deterministic approach. These problems concern the choice of the model complexity level, the coupling, at the interface between two media, of models representative of a same phenomenon, and the optimization of the calculation management.

The French experience allows us to be optimistic in the resolution of the problems.

Résumé

On distingue généralement, pour effectuer les calculs du risque associé aux stockages de déchets une méthodologie "probabiliste", mettant en oeuvre, compte-tenu des possibilités actuelles des ordinateurs, des modèles simples, et une méthodologie "déterministe" capable d'intégrer des modèles détaillés.

L'approche française participe à ces deux méthodologies. Nous décrivons ici, sur la base de l'expérience acquise dans la mise au point du code MELODIE, quelques problèmes à résoudre dans l'intégration de modèles détaillés dans une approche déterministe. Ces problèmes concernent, outre la validation physique des modèles utilisés (problème qui n'est pas traité ici), le choix du niveau de complexité des modèles, le couplage à l'interface entre milieux voisins, de modèles représentatifs d'un même phénomène ainsi que l'optimisation de la gestion des calculs.

L'expérience acquise en France permet d'être optimiste quant à la résolution de ces problèmes.

INTRODUCTION

The safety demonstration of a waste repository rests on the verification of the admissibility of the doses which might be received by the populations in the present time and in the future. Due to the nature and the dimensions of the media involved (the repository, the geological medium, the biosphere) and the time periods considered (from some thousands to hundred thousands of years), this verification is only feasible through calculations.

Two methodologies are usually considered for these calculations (1) :

- The first methodology, called "probabilistic", points out, on the one hand, the high complexity of the disposal media and the variability of their characteristic parameters, and, on the other hand, the diversity and the stochastic nature of the potential future events. This methodology aims at a repository safety's "statistical verification", based on the interpretation of risk diagrams which represent the consequences associated with the repository versus the corresponding scenario frequency (a scenario is defined, in this methodology, as a set of data which characterizes both the initial spatial disposal configuration and future potential events). The implementation of this methodology involves a great deal of calculations, which imply a limitation of the number of phenomena simultaneously taken into account and the use of very simplified models for the description of the phenomena. In some cases, these simplifications are compensated by the widening of some parameter uncertainty curves.
- the second methodology, called "deterministic", aims at obtaining the best possible estimates, for different reasonable sequences of future events (scenarios), of the consequences, in term of doses to the population, of a repository, characterized by a set of initial mean parameters. Due to the lower number of calculations implied by this approach, it is possible to use detailed realistic models. It is important to notice that each best estimate is complemented by an uncertainty margin which takes into account the variability of the physico-chemical characteristics of the media : the methods used to evaluate this margin are usually very similar to those which are used in the probabilistic methodology and, thus, involve simpler models than the best estimate ones.

A PAGIS (2) workshop, held in ISPRA at the beginning of this year, pointed out the complementarity of the two methodologies.

We will present hereafter some reflections, based on the French experience, about the problems encountered in the use of detailed models and their integration in a global model.

THE FRENCH APPROACH

The approach adopted in France by the CEA-IPSN belongs to the second of the previously defined metholologies, and relies on the development of a global model which will progressively take into account the different phenomena involved, after sensitivity studies have demonstrated their influence.

This model, called MELODIE (3), under development since mid-1984, will permit us to perform calculations for the three kinds of geological formations which are considered in France : granite, salt and clay. Its development associates different teams specialized in the various subjects involved. Different exercises, either "analytical", that is, concerning the sub-models, such as the intercomparison exercises INTRACOIN and HYDROCOIN [(4,5)], or global (such as the European PAGIS project), are currently run, in order to validate both the sub-models and the methodology used.

1) Main axis of the development of the model MELODIE

Different stages are foreseen in the development of the MELODIE model.

a) The first stage, currently in progress, consists of the setting up of a version which will permit the best estimate of the transfer of radionuclides in an environment defined by a set of initial data (crossed media characteristics) without any time variation of these data. This version, of which the simplified organigram is given in the figure 1 associates :

- a source model, called CONDIMENT, developed in the Waste R & D Department of the CEA, and presently fitted for glass matrix.

- a geosphere model (hydrological and radionuclides transport), called METIS and developed by the Ecole Nationale Supérieure des Mines de Paris. This model, fitted in its present version for granitic formations, has been involved in the INTRACOIN and HYDROCOIN exercise projects,

- a biosphere model, called ABRICOT, developed by the Technical Protection Department of the CEA.

b) The second stage foresees the addition of a parameter sensitivity studies algorithm. This one, which has been developed in parallel with the first version of the MELODIE model, is currently introduced in it. The corresponding version, of which the organigram is given in figure 2, should be available in the middle of 1986. The method used is the so-called Latin Hypercube Sampling (6). Simplified sub-model versions will be used; in particular, a one-dimensional model for radionuclide transport through the geosphere will be used : the monodimensional pathway (length, physical characteristics) will be determined by the analysis of the bidimensional best estimates.

c) The third stage in the MELODIE development will make it possible to take into account geosphere evolution (modifications of the site's geometry, the medium characteristics, such as the fissuration or the permeability, and the boundary conditions, such as the surface hydraulic head). The phenomena involved are due to the existence of the repository (heat release leading to medium deformation) or to external events (tectonic or climatological events, human intrusion). Preliminary studies will be necessary, for instance :

- the determination of the numerical behaviour of the METIS hydrological sub-model when the geosphere characteristics are time-dependent,

- the choice of a methodology to take into account the mechanical effects associated with the thermal repository load. It is important to recall here

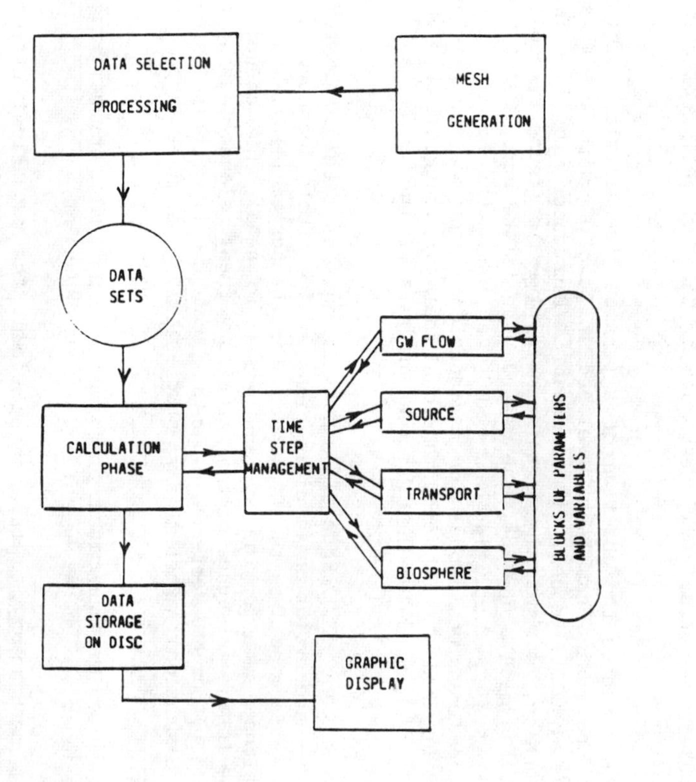

Figure 1

MELODIE 1st VERSION (APRIL 1985)

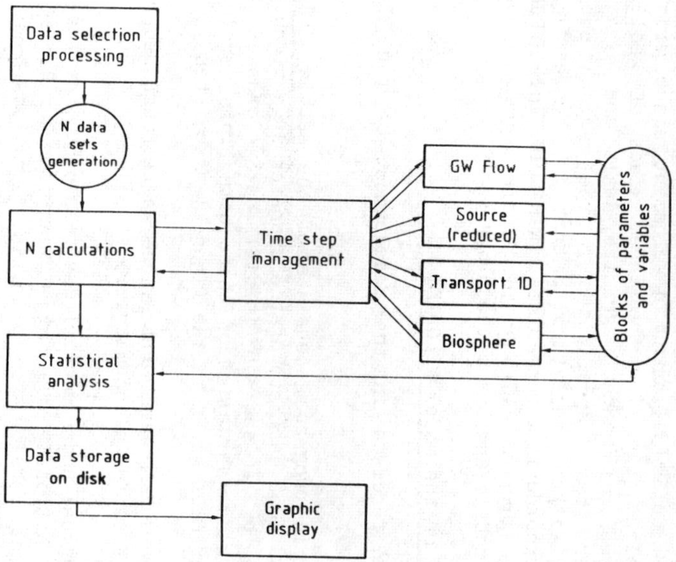

Figure 2

MELODIE : SENSITIVITY ALGORITHM

that, when the risk analysis is performed, the repository has already been dimensioned through thermo-mechanical calculations which make it possible to consider that no inadmissible deformation has occurred : the thermo-mechanical effects are thus supposed to have only a second-order effect on the transfer of radionuclides,

- the definition of scenarios (sequences of events). Geoprospective studies (7), led in consultation with the Bureau de Recherches Geologiques et Minières, will allow us to define the most probable scenarios. These studies will be followed by the development of models able to determine the impact (erosion, geosphere fracturation and permeability modifications,...) of the foreseen events (glaciations, evolution of the sea level, epirogeny,...).

2) Problems encountered in the use of detailed models in the deterministic approach

The first problem is, of course, related to the physical validation of the sub-models used. However, this problem will not be examined here, due to the fact that it will be considered elsewhere in the workshop.

The other problems encountered in the integration of detailed models in a global model are either numerical in nature, or associated with the optimization necessary to avoid untenable computer times. We will develop hereafter three examples, which concern :

- the choice of the level of complexity of the sub-models,
- the coupling, at the interface between two neighbouring media, of models representative of a same phenomenon,
- the optimization of the management of the calculations.

Another important problem will not be treated here. It concerns the coupling, within the same medium, of models representative of different physical phenomena simultaneously involved. Some studies in this field are currently in progress concerning, for instance, the coupling of hydrology and heat transfer (thermo-convection), but they are at the present time performed outside the global model and it would not be possible to draw lessons from them.

2.1 Choice of the level of complexity of the sub models

The choice of the level of complexity of the sub-models is based on certain rules, which are :

- if the involved phenomena are ill-understood, a complex model would be premature, meaningless and sometimes harmful;

- when various phenomena are simultaneously involved, the impact of one of them, in a given domain of parameters, can be negligible compared to the impact of the others; the choice of a complex model, describing all the phenomena, may then be superfluous in the considered domain;

- for a given use (preliminary calculations,...), the desired level of accuracy can be relatively low : a simple model would then be sufficient;

- economic reasons (computer costs) can lead to the preference of a simple model rather than a complex one. It is however noticeable that the notion of complexity in this case, is a strong function of the nature of the tools (hardware and software) used.

For the development of the MELODIE model, the previous reasons justify some present choices. We will give three examples :

- the phenomena involved in the source evolution (ageing and degradation of the materials; radionuclides release and transfer, physico-chemical interactions...) are at present ill-understood. This is the reason why the source sub-model, presently introduced in MELODIE, is rather simple : sudden degradation of the container at a predetermined time, release of the radionuclides controlled either by the diffusion inside the glass matrix or the solubility limit in the water, diffusive transfer through the backfilling material, radionuclides-bentonite chemical interactions described by a linear isotherm. Major effort has been foreseen in this area in order to gain a better understanding. In particular, a radionuclide release model, based on the hypothesis of glass matrix dissolution and the congruent dissolution of the radionuclides, followed by selective reprecipitations, is currently studied. In the first stage, this model will not take into account the effect of the layers created by the precipitates, which is a conservative assumption.

- the use of a two-dimensionnal version of the METIS model for the flow calculations through the geosphere rather than a three-dimensionnal version is due to economic reasons. However, a three-dimensionnal version is available, outside MELODIE, and will be used in order to determine the best position of the 2-D cross-sections,

- the geochemical interaction model introduced in METIS is a simple model (linear isotherm), because it would be illusive, in the present state of the art, to do otherwise. However, a study, led in parallel to the development of MELODIE, aims at a best representation of the geochemical interactions between the transported radionuclides and the geological medium. This study, based on the research of the chemical equilibria between the components in both the solid and liquid phases, will make it possible on the one hand to interpret at best the experimental results and, on the other hand, to avoid the multiplication of experiments to determine the Kd-values which will be introduced in the simple operational model. The studies on this complex model will probably be long and it is not realistic to think, at the present time, that it will be possible to introduce this model in MELODIE, except eventually to simulate the environment and so facilitate the setting up of particular phenomena, or justify some hypothesis.

2.2 Coupling, at the interface between two media, of models representative of the same phenomenon

The mass, momentum and energy transfers between neighbouring media are taken into account in the numerical models by couplings. The present version of MELODIE couples together the radionuclide transport models in the source and in the geosphere. This coupling is interactive, which means that, at each time, that is numerically at each time step, the continuity of the parameters which characterize the radionuclides transfer (flux and concentration) is insured;

Let S be the source model (CONDIMENT) and G the geosphere model (METIS). We can write :

$$\varphi = S(\bar{c}, x)$$
$$c = G(\bar{\varphi}, y)$$

in which

$\bar{c}$ is the concentration which is set, inside the source model, at the interface between the source and the geosphere, and φ is the flux calculated by the source model at this interface,

$\bar{\varphi}$ is the flux at the inlet of the geosphere, which is set in the geosphere model, and c is the concentration calculated by the geosphere model at the interface,

x and y represent the other parameters which are involved in the source and geosphere models.

The coupling conditions are the following :

$$c = \bar{c}$$
$$\varphi = \bar{\varphi}$$

In order to insure the coupling, we have developed an algorithm based on a Newton method and an iterative process.

The interest of an interactive coupling and the problems encountered are illustrated by the first results obtained by the CEA-IPSN in the frame of the PAGIS exercise.

These results concern the Auriat reference site : a two dimensional cross-section, along a SSW/NNE axis, beginning at l'Etang du NOUHAUD (at the boundary of the granite màssif), going through the boreholes and the summit, and ending at 1 km at the NNE of the LA VIGE river, was chosen (see figure 3). The geosphere has been represented by an homogeneous equivalent medium, in which four permeability zones have been distinguished versus the depth. The hydrogeological calculation has made it possible to determine the flow in each point and the position of the resurgencies (figure 4). The repository, situated in the lower permeability zone, at - 675 m deep, and at about 1 km SW from the summit, is described as a succession of elements, each of them being centered on a geosphere mesh node, and corresponding to a determined number of waste packages. Each waste package is modelled by a cylinder constituted by the glass matrix, the surrounding bentonite and a small portion of the host-rock. The radionuclides flux going out of each waste package is calculated at each time step, making it possible to determine the flux going out of each element of the repository.

The first calculations have concerned three beta-gamma emitters radionuclides : Cs-135, Tc-99 and Zr-93.

Figures 5 and 6 give respectively, for Cs-135, the activity flowrates at the different resurgencies and the individual doses for the La Vige resurgency. Each of these figures is composed of both the results obtained when the models are not coupled together (case a) and the results obtained with the interactive coupling (case b).

We can see that :

- two peaks appear on the resurgency activity curves : these peaks have no physical meaning but are due to numerical problems. These problems have now been solved, but we have not yet calculated again the Auriat case.

- the effects of the interactive coupling are first, to delay the time position of the maxima of the activity flowrates and corresponding doses, and secondly, to reduce the value of these maxima by of about 10 times. These effects are well representative of the physical impossibility of the geosphere to "digest" instantaneously the flux of radionuclides which goes out of the source.

2.3 Optimization of the calculation management

The use of detailed models in the deterministic approach presents, a priori, the danger of leading to the development of a very large computer tool which would require much computer time and hence be costly to run, ill-fitted for model evolution and the introduction of new models. That is why the optimization of the calculation management under its numerical as well as its computer related aspects is to be considered very soon.

Apart from the choice of the level of model complexity, this optimization also needs in particular :

- the sub-division of the different tasks involved (data management, calculations chaining, calculations performing, results management,... (see figure 7)), the modularity of the global structure, and the associated writing rules for computational modules. This last point is particularly important in the realization of MELODIE because independant teams are developing the different models,
- the combination, according to the stages of the calculation, of different computers (micro-computers for the elaboration of the data sets and their control tests, CRAY-type computers for the calculations, IBM-type computers for the treatment of results,...) and the optimal use of each of these computers (vectorialization of the algorithms for the calculations on the CRAY, for instance)
- the use of advanced programming languages, and of softwares specially fitted for the memory management or the chaining of calculations,

- the introduction of safeguards-restart procedures. Such procedures allow a better control of the progress of the calculation, and make it possible to modify, if necessary, the models or the data without beginning the whole job again : this possibility gives a high flexibility in order to study the effects of various scenarios,

- the use of different time steps according to the needs of different models. Each sub-model possesses an optimal time step, depending on the time constants of the physical phenomena which are represented and on the numerical schemes. The optimization of the global system must respect these time step differences and chain the sub-systems by taking into account this constraint. In the example presented in the § 2-2, different time steps have been used for the source and the geosphere models.

CONCLUSION

The use of complex models in deterministic evaluations of waste repository performance solves some problems which are either numerical or associated with management optimization considerations.

The French experience in this field allows us to be optimistic. The first calculations performed upon a complex repository configuration with the MELODIE model, which associates interactively a source, a geosphere and a biosphere detailed model, have made it possible to highlight some of these problems and how to solve some of them.

REFERENCES

1 - Long-term risk assessment of geological disposal : methodology and computer codes - G. Bertozzi, M. D. Hill, J. Lewi and R. Storck. Deuxième conférence des Communautés Européennes sur la gestion et l'évacuation des déchets radioactifs (Luxembourg, 22 au 26 avril 1985)

2 - PAGIS (Performance Assessment of Geological Isolation Sytems) - Summary Report of phase 1 - Commission of the European Communities - Rapport EUR 9220 EN.

3 - Development of the code MELODIE for long term risk assessment of nuclear waste repositories. P. Coblet, P. Guetat, J. Lewi, J.P. Mangin, G. de Marsily, P. Raimbault - MRS conference (Stockholm, September, 9-11, 1985)

4 - INTRACOIN - Final report level 1 (February 1984)

5 - HYDROCOIN - Progress report n°1 (May 1984 - December 1984)

6 - R. L. Iman - M. J. Shortencarier - A fortran 77 programm and user's guide for the generation of Latin Hypercube and Random Samples for use with computer models - Technical Report - Sandia 83-2363 SNL - Albuquerque (N.M.)

7 - Rapports CCE/BRGM : étude géoprospective d'un site de stockage (9 documents).

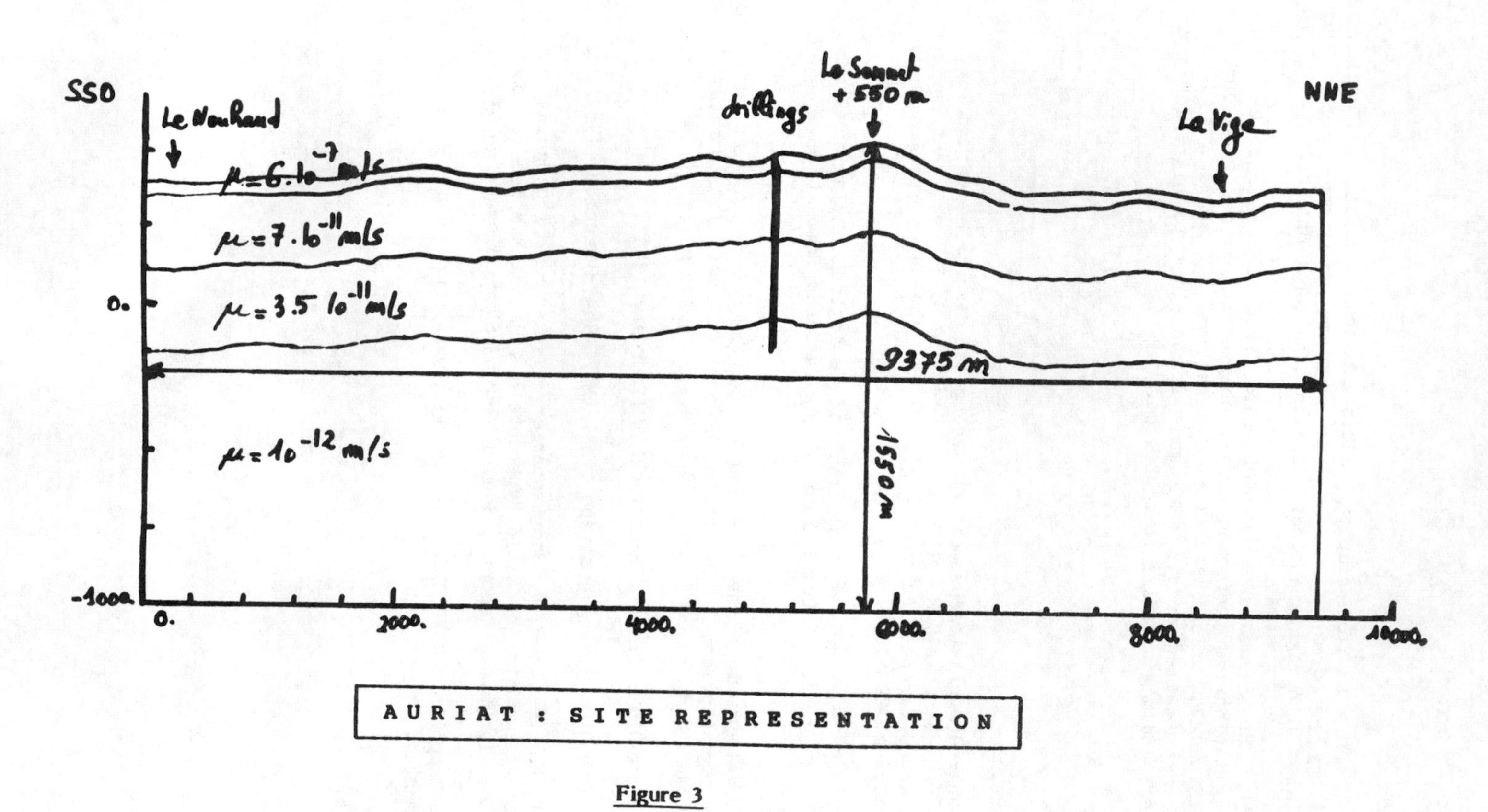

AURIAT : SITE REPRESENTATION

Figure 3

AURIAT . PORE WATER VELOCITY

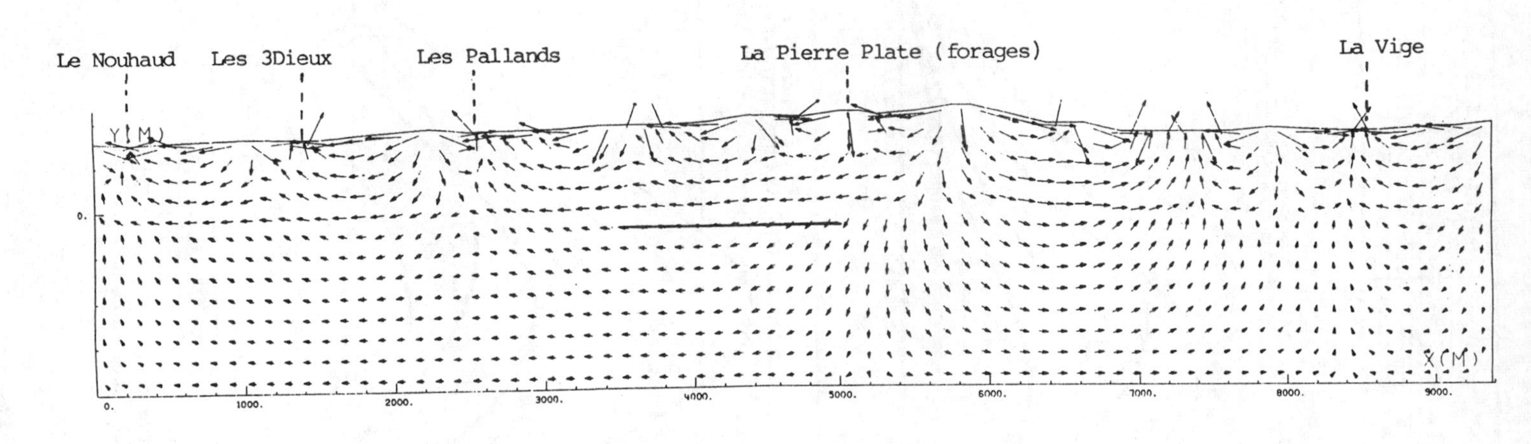

altitude (m)	P.W. velocity (m/year)
470	22
380	$0.14 .10^{-1}$
260	$0.12 .10^{-1}$
140	$0.94 .10^{-2}$
20	$0.91 .10^{-2}$
-100	$0.19 .10^{-2}$ approx. location of repository

altitude (m)	P.W. velocity (m/year)
-200	$0.17 .10^{-2}$
-340	$0.15 .10^{-2}$
-450	$0.13 .10^{-2}$
-600	$0.11 .10^{-2}$
-700	$0.99 .10^{-3}$
-1000	$0.84 .10^{-3}$

Figure 4

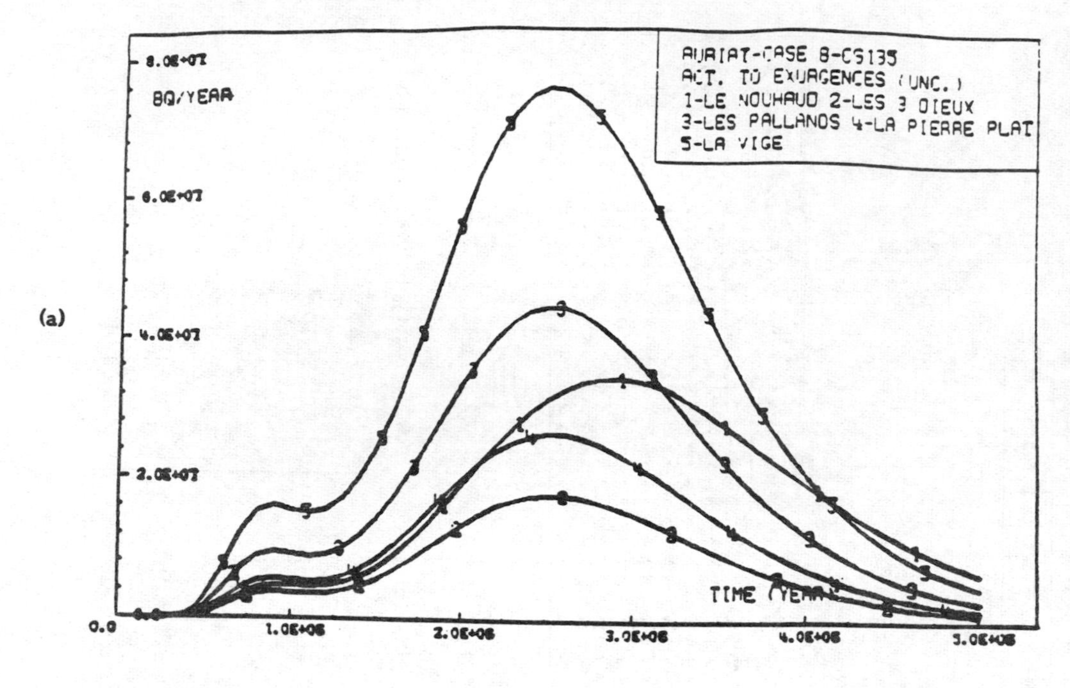

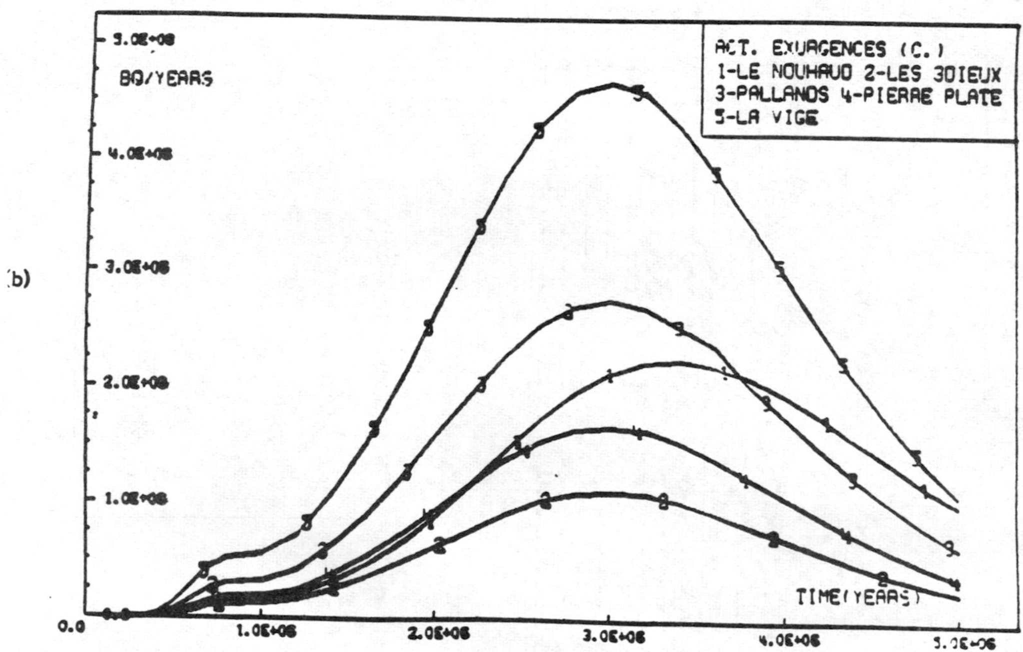

Figure 5 : Cs-135 activity to exurgencies
(a) : uncoupled case (b) : coupled case

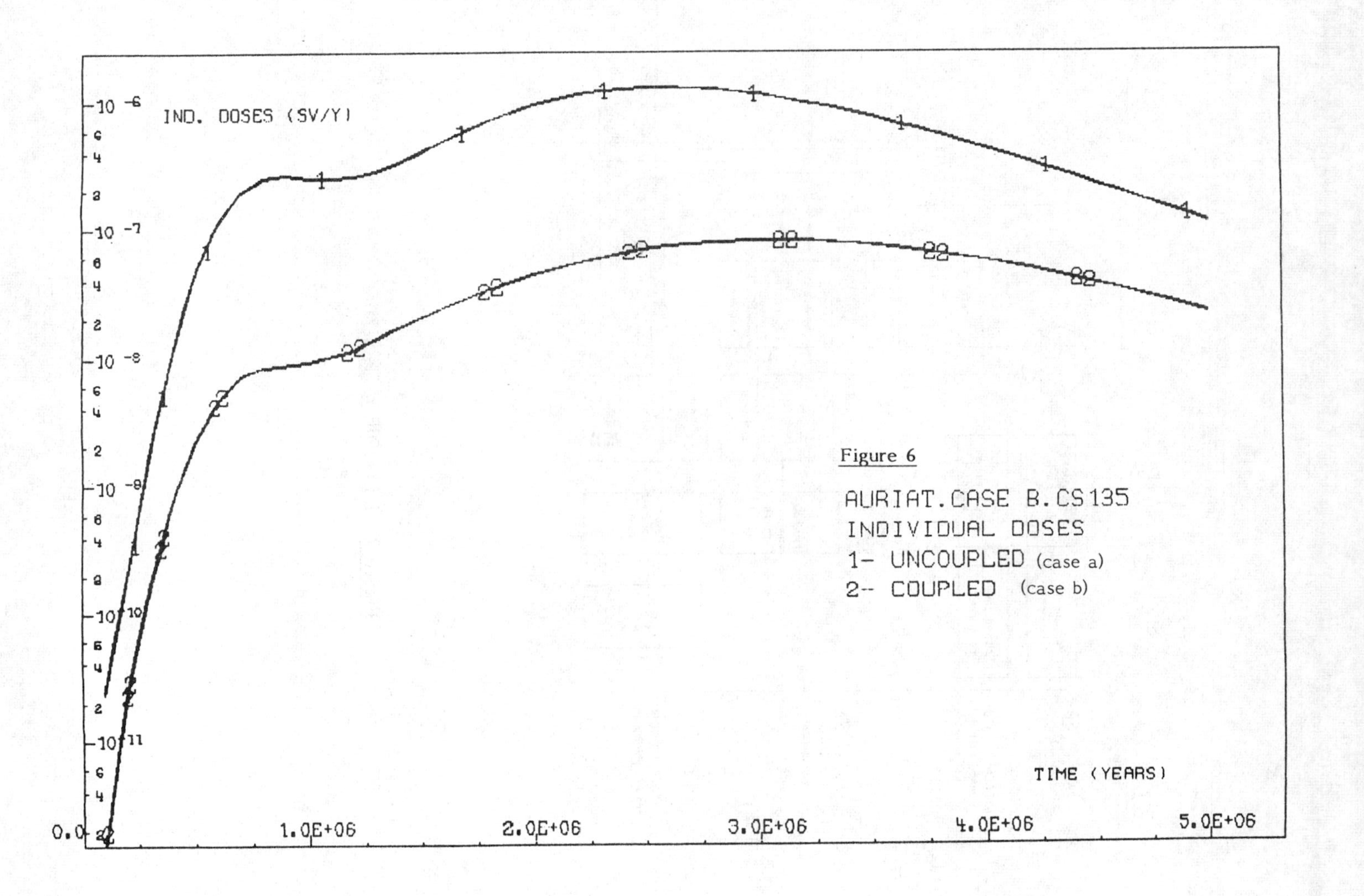

IND. DOSES (SV/Y)
AURIAT.CASE B.CS135
INDIVIDUAL DOSES
1- UNCOUPLED (case a)
2-- COUPLED (case b)
TIME (YEARS)
0.0
1.0E+06
2.0E+06
3.0E+06
4.0E+06
5.0E+06

Figure 6

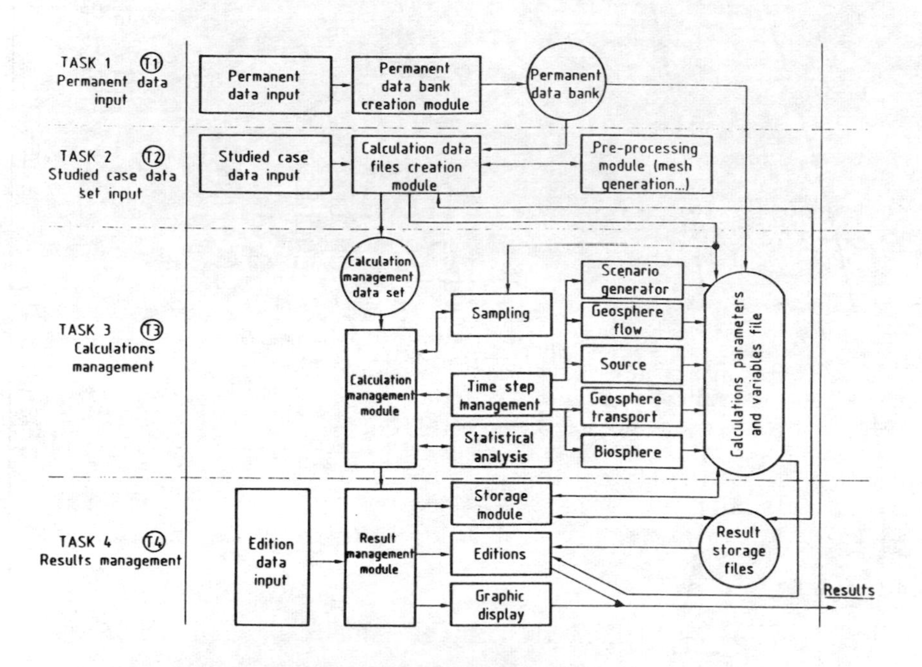

Figure 7 : MELODIE tasks parting

COUPLING OF NEAR-FIELD PROCESSES IN SALT DOME REPOSITORIES FOR RADIOACTIVE WASTES

R. Storck and S. Hossain
Gesellschaft für Strahlen- und Umweltforschung mbH München
Institut für Tieflagerung

ABSTRACT

The various links of the near-field processes integrated into the performance assessments of salt dome repositories in Germany have been described. These links have been established with the help of the computer code EMOS, which is based on compartment model. For the modelling of the physical and chemical processes in each compartment, simplified models have been used. These simplified models have however been derived from the detailed models of the processes with their laboratory and/or in-situ validations.

COUPLAGE DES PROCESSUS EN CHAMP PROCHE DANS LES DEPOTS DE DECHETS RADIOACTIFS AMENAGES DANS DES DOMES DE SEL

RESUME

Les auteurs décrivent les diverses relations entre les processus en champ proche intégrés dans l'évaluation des performances de dépôts aménagés dans des dômes de sel en Allemagne. Ces relations ont été établies au moyen du programme de calcul EMOS, qui est fondé sur un modèle à compartiments. On a utilisé des modèles simplifiés pour modéliser les processus physiques et chimiques à l'intérieur de chaque compartiment. Ces modèles simplifiés reposent toutefois sur les modèles détaillés de processus qui ont été validés par des expériences en laboratoire et/ou sur le terrain.

1. INTRODUCTION

In the long-term safety analysis of a repository for radioactive wastes, links between various physical and chemical processes within the system have to be integrated for an overall performance assessment. Since these links of various processes are complex, simplified mathemetical models characterizing these processes are unavoidable. These simplified models should however be based on detailed models of the processes, which have to conform to field observations and experimental measurements.

How these different links have been achieved in the German approach, will be shown with examples in this paper. At first, an overview of the present status of performance assessment in Germany will be given. In the following detailed description of links between processes, however, we shall concentrate only on those in the near-field of the repository.

2. THE STATUS OF PERFORMANCE ASSESSMENT IN GERMANY

At present, the performance assessment in Germany is based on a deterministic approach, in which the calculation of maximum individual radiation doses for chosen scenarios are carried out. In the case of the salt dome repository, the scenario under study is characterized as follows:

> Intrusion of brine from the overlying rock into the mine workings via the Main Anhydrite at the beginning of the post-operation period.

To calculate the consequences of this scenario, the whole system is split into three subsystems: salt dome, geosphere and biosphere (Figure 1). The section salt dome comprises the disposed wastes as well as the possible mobilization and the transport of radionuclides to the boundary of the salt dome. The section geosphere deals with the transport of released radionuclides along the path of groundwater flow and a possible contamination of the groundwater near the surface. The section biosphere is concerned with different utilizations of groundwater and resulting radiological consequences to man.

For the modelling of these subsystems, three computer codes are available. The release from the repository is calculated by the compartment model EMOS, which has been developed in our group. The structure of this code has been described in [1]. For the transport of radionuclide through geosphere, the three-dimensional finite-difference code of Intera Environmental Consultants SWIFT [2] has been used. The modelling of biosphere to calculate the potential radiation exposure has been done by the code ECOSYS, which is also a compartment model developed in GSF München [3].

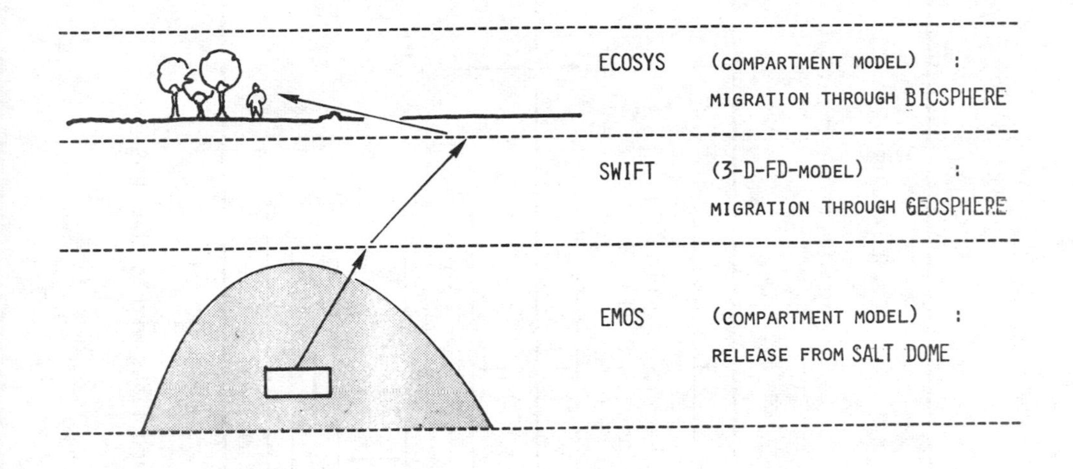

Figure 1. Schematic representation of the three subsystems for the performance assessments of a salt dome repostory in Germany

3. NEAR-FIELD MODELLING

A. Structure of the Model

As already mentioned in the previous Section, for the modelling of the transport of radionuclides through the backfilled repository and subsequently the release of radionuclides at the boundary of the salt dome (interface with the geosphere), the compartment code EMOS has been developed. In Figure 2, the structure of different levels and links within this code has been shown.

In the first level, general executives such as the preparation of input-/output-data, the management of data base etc. are accomodated. These are closly linked with the second level containing the compartment models, each of which represents a part of the repository subsystem such as one waste package, one borehole for HLW, etc. These compartments, also called barriers, contain further executives for all the processes which have to be considered within this part of the repository. The third level contains the simplified models of different phenomena which are interconnected with each other and called from the compartment executives.

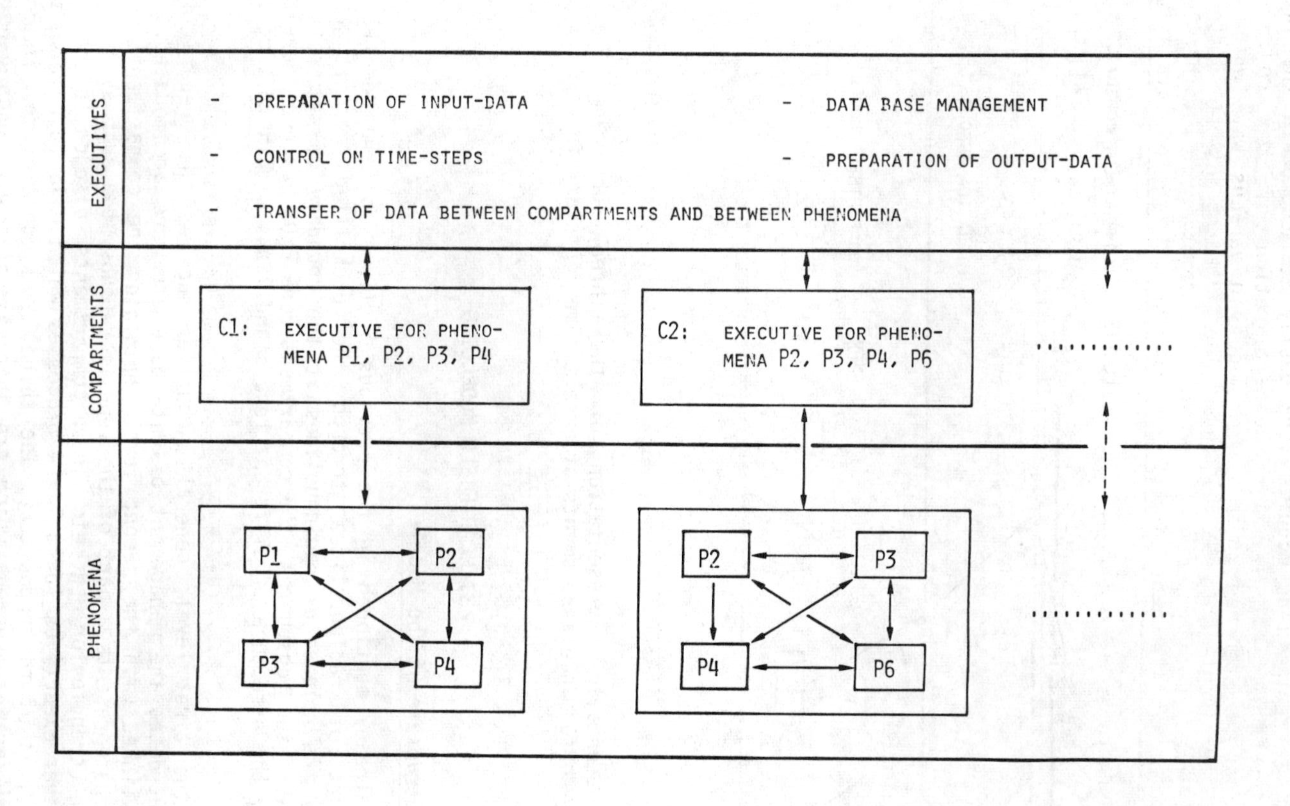

Figure 2. The structure of different levels and links within the EMOS-code

During the process of development of the code, all compartments and phenomena can be replaced as modules very easily to accomodate the latest status of research findings. For the application of the code, the compartments can be combined to a system model in various ways which are provided in the input-data of the code. The linkage between different processes are handled either directly within one compartment or indirectly between processes of different compartments via the transfer of data between them.

B. EXAMPLES OF COMPARTMENTS AND PROCESSES

In the near-field modelling, the subsystem repository has been divided into different compartments, through which the radionuclides are transported successively. According to the nature of the processes occuring in these compartments, they can be grouped in the following main types:

- waste package for HLW (glas)
- waste package for MLW/LLW (concrete)
- sealed boreholes
- backfilled and sealed chambers
- backfilled galleries

As mentined above, these compartment models consist of links between processes to be considered in these compartments. The main processes considered in the assessment of a salt dome are as follows:

- container failure
- leaching from waste matrix
- convergence caused by creep of the rock
- forced convection in networks of gallery system
- natural convection
- nuclide retention by sorption and precipitation
- nuclide transport by forced and natural convection
- decay of radionuclides

The detailed description of these compartments and phenomena are given in [1].

4. LINKS BETWEEN PROCESSES IN A COMPARTMENT

A compartment model has the task of linking all the processes to be considered within that compartment. How such a linkage is established will now be shown with a concrete example. For this purpose, we choose the compartment of backfilled gallery. However, for simplicity we only consider

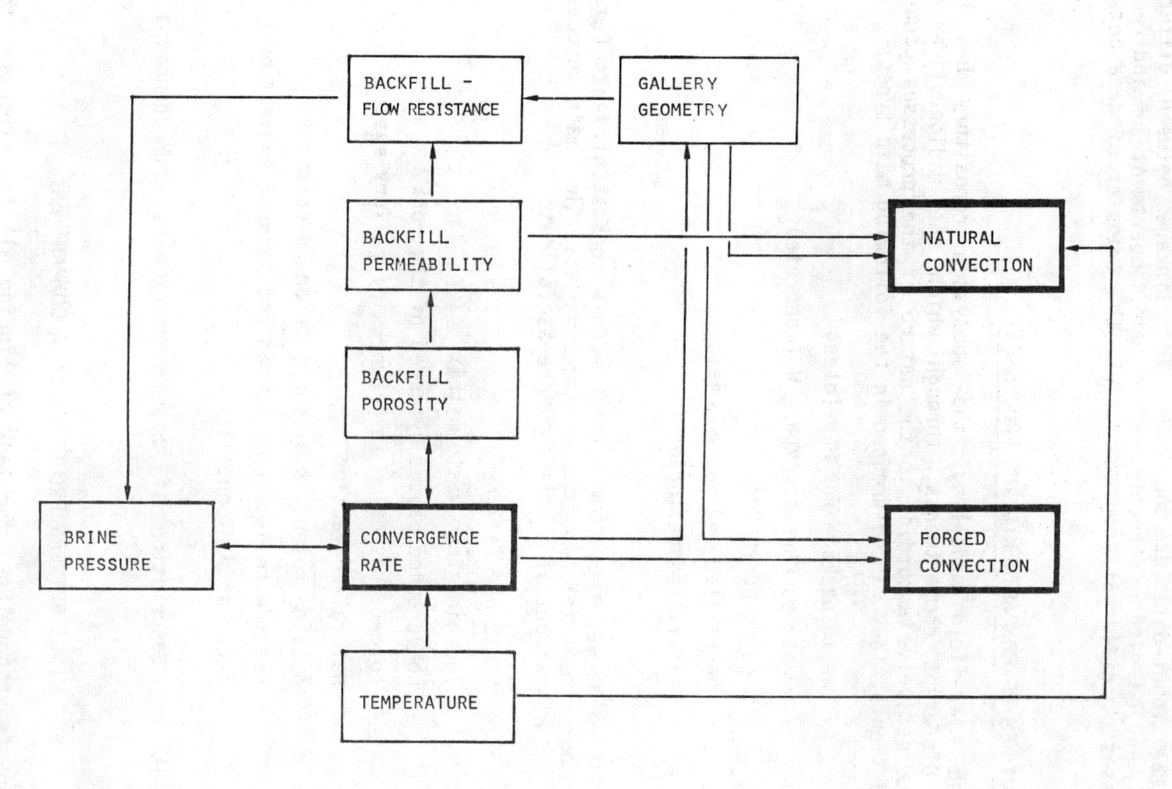

Figure 3. Links between the physical phenomena in the compartment of the backfilled gallery

the physical processes such as convergence rate of the backfilled gallery, and forced and natural convection, which represent only about 30% of the total links within this compartment. The transport of radionuclides, the retention of radionuclides by sorption, etc. have not been considered.

As shown in Figure 3, the convergence of the backfilled gallery due to creep of the salt influences its geometry, the porosity of the backfill within it and the brine pressure. The convergence rate itself is influenced by the temperature, the porosity of the backfill and the brine pressure. The change in the porosity of the backfill material changes its permeability, which when combined with its geometry determines the resistance of flow through the backfilled gallery. This resistance together with the convergence rate controls the brine pressure. The processes which are directly involved in the transport of radionuclides are forced and natural convection. The forced convection is affected by the convergence rate and the geometry of the gallery. When only the thermal effect is considered, the natural convection depends on the temperature, the geometry of the gallery and the permeability of the backfill.

5. ESTABLISHMENT OF SIMPLIFIED MODELS FOR PROCESSES

In this Section, it will be shown how a simplified model to be used in the near-field modelling is derived. For this purpose, we consider the special case of the convergence rate, the simplified model of which is based on both the detailed numerical calculations and the laboratory/in-situ experiments. Figure 4 shows these links between different efforts, which contribute towards the final simplified model of the convergence process.

Detailed numerical calculations [4,5] on the convergence or creep of the rock salt are based on creep laws which are validated by laboratory and in-situ experiments. The simplified modelling has been done via analytical considerations in the whole range of the parameters, temperature (T), fluid pressure (p) and porosity (n) of the backfill, influencing the convergence process. In this formulation, the reference convergence rate K_{ref} and the parameters for the functions $f_1(T)$, $f_2(p)$ and $f_3(n)$ are still to be validated. The parameters in functions f_1, f_2 and f_3 have been validated using the numerical calculations in some range of temperature, pressure and porosity. Finally, the reference convergence rate has been derived using experts' opinion based on in-situ experiments at p=n=0 and long-term field observations, for which $T=T_G$, $n>0$ and p=0.

6. CONCLUDING REMARKS

In this report, it has been shown how complex is the coupling between different processes in the near-field of a repository. Thus, for proper understanding of the behaviour of individual processes, the modelling should be on the one hand as simple as possible and on the other hand include all the importants effects within them. Since the individual simplified models will remain in a continual development process, the computer code integrating these processes should be flexible enough to accomodate the current status of development.

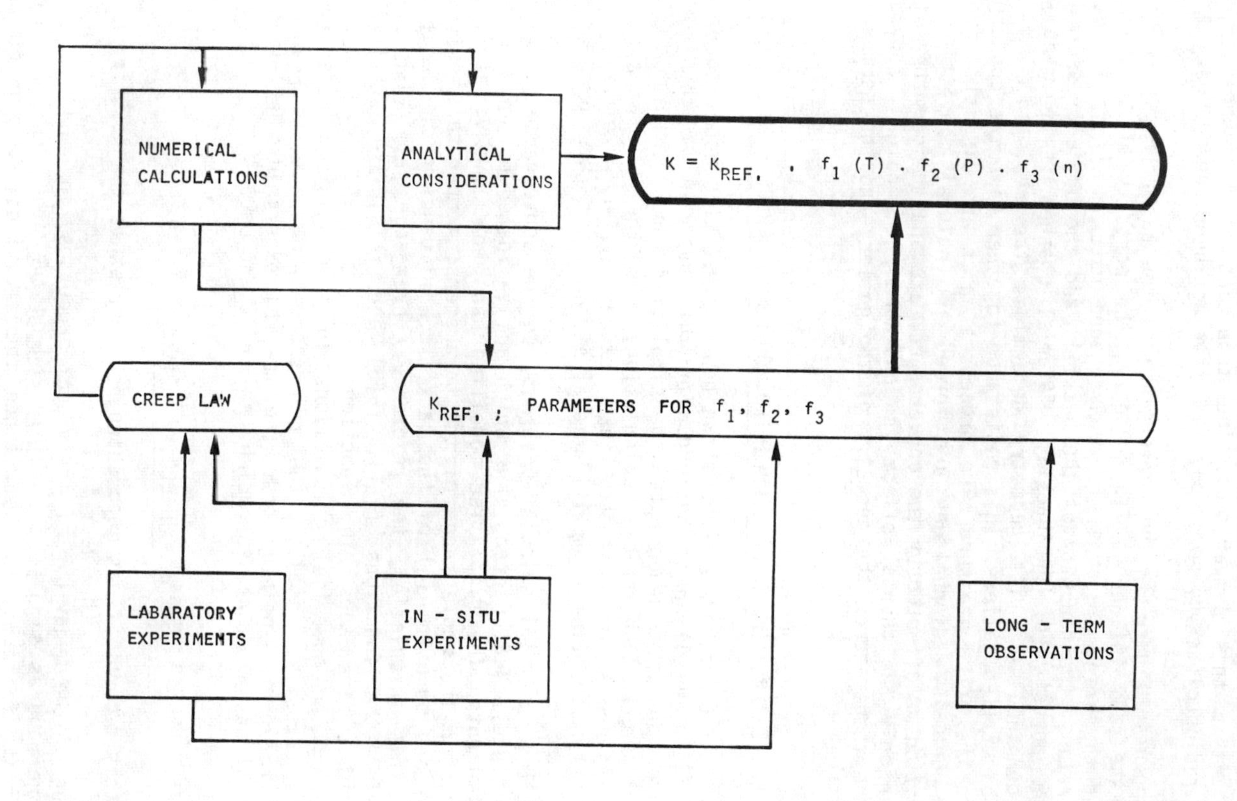

Figure 4. Establishment of a simplified model for the convergence rate

REFERENCES

1. Projekt Sicherheitsstudien Entsorgung (PSE), "Modellansätze und Ergebnisse zur Radionuklidfreisetzung aus einem Modellsalzstock", Abschlußbericht, Fachband 16, PSE, Berlin, 1985

2. Intera Environmental Consultants, "The Intera Simulator for Waste Injection, Flow and Transport - User's Manual for SWIFT", Intera Environmental Consultants, Inc., Houston, 1982

3. Projekt Sicherheitsstudien Entsorgung (PSE), "Potentielle Strahlenexposition durch Nutzung von radioaktivkontaminiertem Grundwasser", Abschlußbericht, Fachband 19, PSE, Berlin, 1985

4. Projekt Sicherheitsstudien Entsorgung (PSE), "Standsicherheitskriterien für das Endlagerbergwerk Gorleben", Abschlußbericht, Fachband 11, Berlin, 1985

5. Projekt Sicherheitsstudien Entsorgung (PSE), "Untersuchungen zum Nahbereich Abfallprodukte/Behälter/Einlagerungsorte", Abschlußbericht, Fachband 12, Berlin, 1985

SESSION IV

THE LINK BETWEEN MODEL DEVELOPMENT AND FIELD/LABORATORY OBSERVATIONS

RELATION ENTRE LES OBSERVATIONS EN LABORATOIRE/SUR LE TERRAIN ET LES MODELES

C.R. Cole and M.G. Foley
Pacific Northwest Laboratory
Richland, Washington U.S.A.

RESUME

Il importe de comprendre, dans les évaluations des performances des systèmes, les diverses relations autour desquelles s'articulent les éléments constitutifs des programmes d'évacuation. Les auteurs définissent la relation entre l'élaboration de modèles et les observations sur le terrain/en laboratoire comme étant établie par le programme itératif de caractérisation des sites et des systèmes qui permet de constituer une base de données d'observation et de vérification. Cette base de données est conçue pour élaborer, affiner et étayer des modèles conceptuels simulant le comportement des sites et des systèmes. Le programme consiste à recueillir des données et à exécuter des expériences propres à démontrer que l'on appréhende les phénomènes intervenant à diverses échelles de temps et d'espace et présentant plus ou moins de complexité. L'aptitude à comprendre et à prendre en compte l'augmentation des incertitudes liées à la caractérisation qu'entraîne le passage à des échelles supérieures de temps et d'espace constitue un aspect important de la relation entre les modèles et les observations. Le processus de fixation des performances (objectifs et niveaux de confiance), associé à une méthode d'évaluation des performances (estimation de ces performances et niveaux de confiance), déterminera le moment auquel la caractérisation aura été suffisamment poussée. A chaque itération, les objectifs de fixation des performances sont examinés et révisés, le cas échéant. On utilise la base de données mise à jour, ainsi que des moyens et méthodes appropriés d'évaluation des performances, afin de recenser et de concevoir les essais et données complémentaires nécessaires pour atteindre les objectifs présentement visés en matière de fixation des performances.

THE LINK BETWEEN LABORATORY/FIELD OBSERVATIONS AND MODELS

C.R. Cole and M.G. Foley
Pacific Northwest Laboratory
Richland, Washington U.S.A.

ABSTRACT

The various linkages in system performance assessments that integrate disposal program elements must be understood. The linkage between model development and field/laboratory observations is described as the iterative program of site and system characterization for development of an observational-confirmatory data base. This data base is designed to develop, improve, and support conceptual models for site and system behavior. The program consists of data gathering and experiments to demonstrate understanding at various spatial and time scales and degrees of complexity. Understanding and accounting for the decreasing characterization certainty that arises with increasing space and time scales is an important aspect of the link between models and observations. The performance allocation process for setting performance goals and confidence levels, coupled with a performance assessment approach that provides these performance and confidence estimates, will determine when sufficient characterization has been achieved. At each iteration, performance allocation goals are reviewed and revised as necessary. The updated data base and appropriate performance assessment tools and approaches are utilized to identify and design additional tests and data needs necessary to meet current performance allocation goals.

1. INTRODUCTION

The importance of performance assessment techniques in judging the safety of radioactive waste disposal has brought us together at this conference. Quoting from the stated objectives of this NEA project:

"The focus of the project should clearly be on current problems in carrying out assessments of performance of waste disposal systems. The increasing number of methodologies, models and criteria has led to some confusion on how such assessments should be carried out. It is suggested that there is one particular problem which requires resolution, i.e., how to rationalise all the various elements that combine together in carrying out performance assessments. This rationalisation is necessary in order **to generate confidence that performance assessments are reliable, realistic and can provide the correct type of information on which the efficacy of disposal can be judged.** Fundamental to this rationalisation to breed confidence is an awareness of the various links between each component of the system. Different types of linkage exist such as the ones between those utilizing and those carrying out performance assessments, through the link between those acquiring data and those utilizing the data in models, to the way individual component models may be coupled together."

Figure 1 shows the linkages in system performance assessments and illustrates the important role they play in judging the adequacy of potential repository sites. The principal linkages are the focus of this conference and include the linkages between:

(1) Output from performance assessments and the needs of regulators
(2) Various predictive mathematical models used in a complete performance assessment
(3) Model development and field/laboratory observations
- **Availability of data; adapting models to data availability?**
- **Use of field data in models and planning data acquisition programs.**

This paper reviews the third linkage area.

2. DISCUSSION

We seek the ability to estimate future behavior with our performance assessment models with as much realism and reliability as is achievable. We do not simply want to **forecast or extrapolate** into the future based on measured data and regression analysis. To provide the predictive capability needed for our assessments we must have a sufficient:

(1) understanding of the processes involved
(2) characterization of the system
(3) sound theory that relates the understanding and characterization
(4) data base of supporting experimental evidence.

The quality we would like to achieve in the predictions made with our performance assessment models was discussed in the Performance Assessment National Review Group (PANRG) report [1] as the important concepts of predictive reliability and bounding analysis. Bounding analysis is one means of achieving the goal of predictive reliability. Predictive reliability is that quality in a performance assessment that ensures the actual performance of a system being studied is as good as or better than the predicted performance [1].

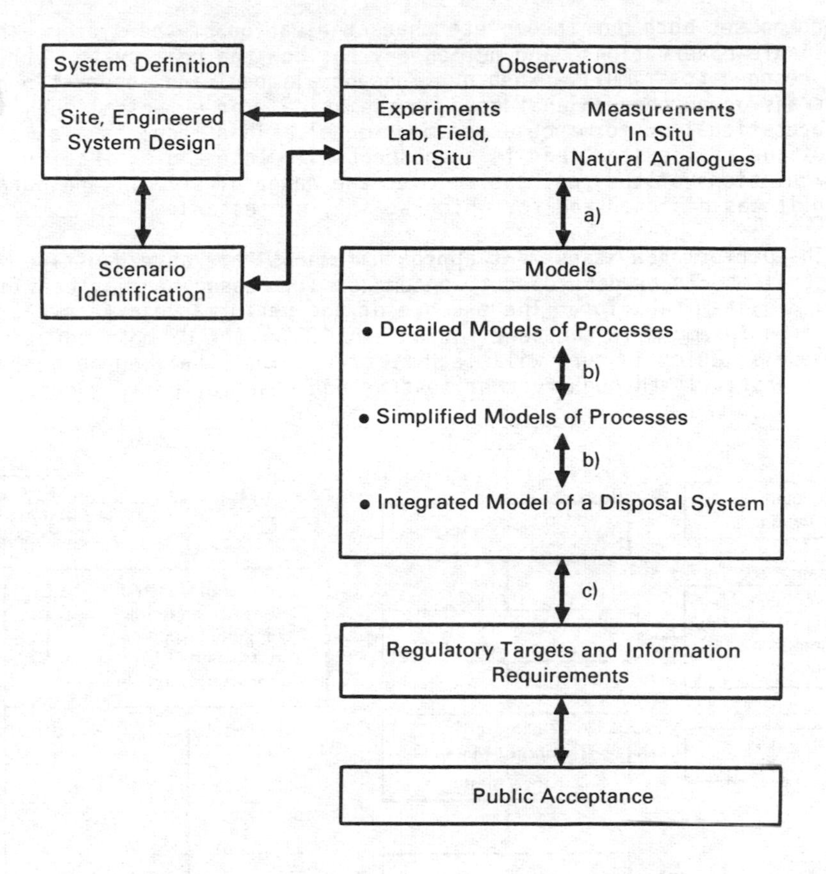

Figure 1. Linkages In System Performance Assessments. (Taken from Note of Meeting of Consultant Group on System Performance Assessment, Paris, 29th and 30th April 1985.)

A. Performance Assessment Modeling

Performance assessment modeling is accomplished by an iterative process of conceptual model development and improvement through a program of testing and response observation. In performance assessment, we generally deal not with simple systems but with complex systems, and complicated conceptual models for describing complicated system responses to a complicated set of stimuli. In performance assessment, the complexity of the systems being studied requires that the modeled system be simplified. The conceptual model encompasses both our incomplete characterization of the system, inferred from available observations, and our theory for how the real system will function and respond to stimuli. When our conceptual model and observational data base is transformed through analogous, mathematical, or numerical methods, we have an operational performance assessment model. This model is a simplified version of the real system, and is supposed to simulate the excitation-response relations of the real system over the range of stimuli and parameters for which it was designed and for which we are interested.

The performance assessment approach demands that an exhaustive set of conceptual models be developed to encompass all reasonable explanations of observed system behavior. The essence of the performance assessment approach lies foremost in the need for an exhaustive set of meaningful conceptual models, which in turn will lead to predictions that can be tested and improved iteratively through further testing and observation (Figure 2). The

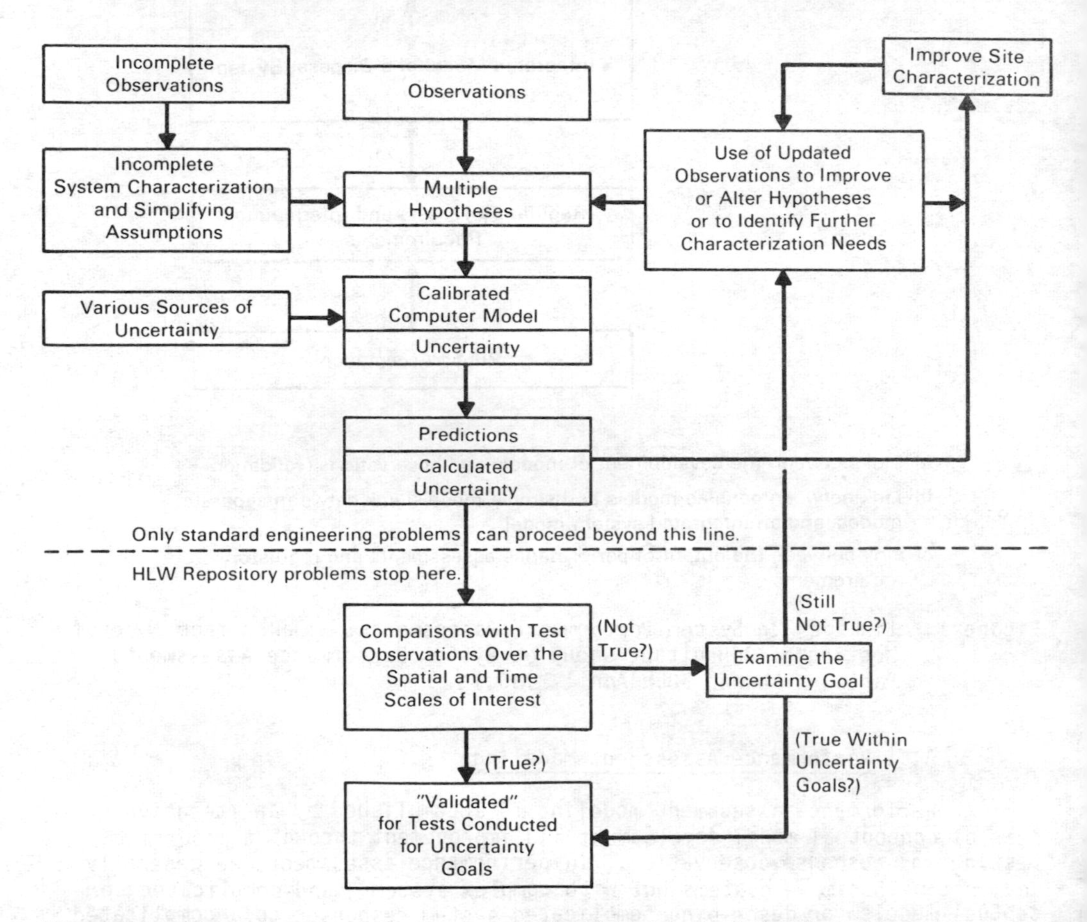

Figure 2. Steps Involved in Traditional Validation of Complex Natural and Engineered Systems.

conceptual model(s) that survives comparison with carefully conceived and controlled experiments, and that withstands peer review by the relevant technical community, achieves the kind of probational acceptance desired to justify its use as a predictive tool.

B. Effects of Scale

In conventional engineering environments, system design lifetimes are a few decades or less so that confirmatory real-time or accelerated time-scaled experiments can be performed to reinforce probational acceptance (i.e., traditional validation of conceptual and performance assessment models). For conventional time scales, historical records for constructing well-documented natural and man-made analogues can be found and used to make objective estimates of long-term performance, and to provide an indirect validation of our understanding of the system. When making performance assessments for thousands to tens of thousands of years, we find that the sheer magnitude of such a time scale creates difficulties with many aspects of the assessment of performance. The effect that previously ignorable chronic phenomena and low-probability events might have on predicted performance must be identified and carefully evaluated. **The integrated effects of chronic phenomena, acting for hundreds of years to millennia, and/or low probability events may destroy many of our usual assumptions regarding time invariability of the parameters describing the system and its performance. We must be aware that our conceptual model of the repository system and site is also susceptible to change as a result of these interacting time-integrated effects, and the effects of low-probability events on the future state of the system (the scenario identification block in Figure 1).** The effect of a changing conceptual model resulting from a consideration of realistic scenarios needs to be accounted for when developing the models and planning the laboratory and field experiments. Due to the expanded time scale for nuclear waste disposal performance assessments, great caution must be exercised when identifying and justifying routine assumptions. In addition to time scale, the spatial scale of interest for our performance assessments, coupled with the inherent spatial variability of geologic systems, leads to the important performance assessment and validation implication of **incomplete characterization.**

C. Uncertainty

All of the effects of time and spatial scale discussed above (no prototype testing, effects of chronic processes on parameters and conceptual models, and incomplete characterization) contribute to uncertainty in performance assessment predictions. Uncertainty in the predicted performance, however, arises from a much broader set of roots, and affects performance assessments of all the components of our disposal system, from the waste form through the canisters, repository, and the site. A few additional sources of uncertainty arise from:

- measurement instrument error
- applicability of the measurement methods and instruments
- parameter interpretation error for inversely determined parameters
- parameter extrapolation error (both time and space)
- site characterization data base
- conceptual model
- scenario identification

- human error
- understanding the application of the basic physics and chemistry (the equations).

A major part of the effort in understanding the important linkages and developing confidence in the reliability of our predictions is going to be concerned with the process of reducing, quantifying, or bounding the uncertainty associated with all the important components required to make performance assessment predictions.

D. Nature of the Linkage Between Observations and Models

Iterative describes the nature of the linkage between observations and models. We start with available system characterization data and observations of system response, then develop preliminary models for use in setting initial performance allocation goals [2] and designing further experiments. The experiments are designed to select between alternative conceptual models and improve the site characterization data base. The purpose of experimentation changes slowly as these early experiments are completed and we gain confidence in our understanding of the physics, the characterization data base, the appropriateness of the conceptual model(s), and our performance allocations goals. The purpose becomes one of developing the observational data base that supports the validity of our detailed subsystem models for the range of conditions that are important to our performance assessment needs, and of determining when we have achieved our performance goals.

Making reliable predictions requires an understanding of the chemistry and physics and the effects of limited characterization at the spatial and time scale important to estimating consequence. Demonstrating understanding at the consequence scale will thus involve experiments at various spatial and time scales and degrees of complexity (more interacting phenomena). As a result, these experiments will range from well characterized to incompletely characterized, and the linkage between observations and models is thus **hierarchical** in terms of scales and complexity. Laboratory and small-scale field (in situ) experiments will demonstrate our understanding of the important basic physics and chemistry at limited spatial and time scales. The models and experiments at this level, while not necessarily simple, are generally much more controlled, and hopefully are well-characterized systems, both spatially and temporally. They will also involve fewer stimuli and basic mechanisms. The experiments conducted on a larger scale are selected to test our theories for integrating the effects of the small-scale phenomena so we can make predictions at this larger scale. It is important to realize that the selection of scale and model can influence not only what parameters are measured, but how they are measured. **There must be a marriage between the theory, for the scale of the test, the observations or measurements that are made, and the tools or instruments used to make the measurements of system responses, system perturbations, and model parameters.**

We all realize that the variety of test scales in terms of time is ultimately limited. We must also realize that **as we proceed to larger and larger spatial scales for our experiments in geologic media, our spatial sampling density and therefore characterization certainty is necessarily reduced. What are the implications of this decreasing certainty in characterization?** One implication is that a series of experiments at these larger scales may be necessary. A set (or sets) of model calibration experiments may be required to allow inverse modeling to be performed to fill in gaps in system characterization parameters. These would be followed by a set of experiments to allow the predictive capability of the calibrated model to be demonstrated.

E. Availability of Data, Adapting Models to Data Availability?

Available data in the form of measurements of parameters, system characterization data, and observations of perturbing stimuli and responses should be used to help us form our initial conceptual model(s) of the system consistent with the performance assessment objectives. The conceptual models are critical in identification and development of data-gathering activities (i.e., site characterization) as well as field and laboratory experiments. The process of attaining the final objectives (i.e., sufficient understanding of the system to make the desired performance assessment predictions) is an iterative process that proceeds from conceptual model identification and selection through data base improvement such that it is "judged good enough" to make realistic or bounding estimates of system performance. At the later stages of this iterative process, further field experiments and data-gathering activities can be used for a dual purpose. First, they can be used to increase our understanding of the system and the system characterization data base. Second, if existing conceptual and numerical models are used to predict the expected results from these new activities in advance, they can also serve to provide confidence that system understanding and the performance assessment model is correct (in some qualitative if not quantitative sense) for the phenomena being tested and the spatial and temporal scale of the test. **It is important to realize that the conceptual model must be clearly identified at every stage of the effort and constantly revisited to check for consistency with the observations.**

As previously stated, prediction requires that for the purposes of our assessment we have a sufficient understanding of the processes involved, characterization of the system, a sufficiently sound theory that relates the understanding and characterization together, and supporting experimental evidence. **Certainly the performance assessment approach (e.g., level of sophistication for the model or bounding analysis) can be adapted to available data, or conservative assumptions can be made to fill data gaps in sophisticated as well as simple models.** Available data can and should be used to help select the conceptual model, but this in turn must be carefully tested. When developing a new theory or testing old theories, one can and should use statistical correlations between available parameters, observations, and stimuli to aid in developing an understanding of relationships. However, these statistical correlations between data sets should not be used as predictive performance assessment models without supporting scientific theory. **In this sense, predictive performance assessment models should not be adapted to available data.**

F. Use of Field Data in Models - Planning Data Acquisition Programs

Previously existing field data should be used **carefully.** The appropriateness (need for adjustment or reinterpretation of raw data, setting of measurement error bands) of existing data for current modeling efforts should be judged by comparison with a well-designed data acquisition program to meet the current system characterization and modeling needs. This does not mean that all existing data which does not conform to current design specifications is useless, but it does allow the importance of the matching or marriage between and among the following to be recognized and properly accounted for:

- purpose of the assessment and stage of the assessment program
- conceptual model or models for the system
- performance assessment approach (e.g., detailed or bounding) and the theory associated with this approach
- scale or sampling size and frequency of sampling in both space and time
- kinds of tools or instruments used to gather the data or make observations
- methods used to interpret and extrapolate these measurements or data.

Let us examine a few hypothetical examples to illustrate the need for caution in using field data, and pose a few questions which should be answered by the field and modeling staff regarding the planning of data acquisition programs.

F.1 Purpose of assessment and stage of assessment program. The purpose of the assessment is to help decide what stimulus-response relationships to capture and the scale (spatial, time, and complexity) to be used. These in turn affect others: conceptual model (degree of simplification), type of analysis, scale and etc. The **bias** that the purpose of previous data collection efforts injects into a data set is often overlooked. Consider an investigation of a multilayered system with a wealth of data from water supply well records. Suppose also the data base is sufficient to give equivalent areal coverage of all layers. Now examine the potential bias. Only the higher permeability zones would be screened and tested, which would greatly affect our statistical sample of aquifer permeability. No permeability determinations would be available for the aquitards. Bias can also result from the purpose of the previous effort coupled with the measurement technique common to investigations with that purpose. Oil well exploration data provides a convenient source of existing data on deep systems but the drill stem testing techniques commonly used to measure hydraulic properties have the potential for routinely excluding any of the higher permeability measurements because of the limitations of the technique. Identification of bias is important to both the use of existing data in models and the design of new data collection efforts.

The stage of an assessment program affects and is important to the design and sequencing of data acquisition activities. Early stage efforts are concerned with identification of the system. Identification information allows us to winnow the various possible conceptual models and thereby plan new, more relevant data acquisition activities (e.g., tests to show whether or not fracture flow is important; data to support or refute steady-state assumptions). Middle stage activities are directed at obtaining the specific relevant parameters and observational data base to support our set of conceptual models, and the later stages are strictly aimed at increasing the observational and parameterization data base to provide whatever is obtainable in terms of validation information to support our system identification, conceptual model, and assessments.

F.2 Conceptual model for the system. The conceptual model is one of the most important aspects of performance assessment modeling and a critical item to the planning of a data acquisition program. A conceptual model is instrumental in determining what needs to be measured. The conceptual model(s), a simple analytical or computer model(s), and some form of sensitivity analyses are required to design site characterization activities. Selection of pumps and other appropriate hardware, measurement instrument design and accuracy, location of observation horizons, and sampling frequency, for example, all require and rely on the conceptual and some simple if not detailed modeling activities. It is important to identify and clearly document the conceptual model or models that were the basis for these experimental designs and to identify the critical assumptions for each possible conceptual model so additional data can be gathered to support or refute the assumptions that are critical to or can differentiate between the different conceptual models.

The importance of the conceptual model to the field and data-gathering activities to the design of appropriate field testing plans is illustrated in the recent paper by Grisak [3]. Grisak discusses how an error in our conceptual model regarding the assumption of a single-density fluid can affect our ability to understand and interpret the results for hydraulic conductivity tests if variable density is present. A field program operating with an inappropriate conceptual model may not include sufficient temperature and total dissolved solids measurements required for interpretation of observations. When the conceptual model accounts for the existence of a vertical temperature or brine density gradient, then these anomalies can be understood and accounted for [3]. Other field problems that arise in the repository program are also related to conceptual model problems, but these conceptual model problems deal with the actual measuring equipment. These include the need for a more detailed model of equipment compliance related effects. In near-surface measuring environments these may not be problematic, but at the depths of repository exploration activities they are amplified and can be problematic [3]. Another problem that arises is related to our desire to infer direction and magnitude of vertical groundwater movement from measurements corrected to "equivalent fresh-water head" when there are both hydraulic and density (temperature and/or brine gradients) related driving forces. Even simple corrections such as borehole deviation must be carefully measured and accounted for due to the great depths of the holes. It is especially critical for all of our standard near-surface measurement techniques to be carefully evaluated in terms of use at the great depths and for the ranges of parameters required for repository characterization programs.

The conceptual model of a geohydrologic system, no matter how technically complex, will always be a simplified picture of the real system [4,5]. Current computer technology and even more importantly data-gathering capabilities simply do not and will not allow a real geohydrologic system to be described in every detail. In some respects modeling a geohydrologic system is an art form [4]. Conceptual model development involves forming a sufficiently accurate simplified picture of the aspects of the system important to making the desired performance assessments. In the process of forming this sufficiently accurate simplified picture, certain groundwater flow and transport modeling technical issues must be considered and addressed, and these decisions, supporting reasoning, and observations must be documented.

The technical issues are simply questions as to what constitutes the correct way to describe the modeled system in terms of relevant processes, parameterization, and numerical models (e.g., saturated or unsaturated flow, buoyancy, importance and method for describing dispersion, importance and required dimensionality, etc.) [4]. The issues stem from limitations on current physical and chemical theories, data-gathering capabilities, and computer modeling capabilities. In some cases these technical issues may not be resolved absolutely.

It is important to realize that the formation of a conceptual model involves both the use of general principles and site-specific data. To illustrate the concept let us walk through an example step that one might take in the development of an initial conceptual model and how this would then need to be supported or refuted by field data collection efforts. An important aspect to the development of a regional model is identification of system boundaries. Examples of general principles that might be used when sufficient data are not currently available would be: (1) regional model boundaries correspond to the topographic highs which separate major surface water drainage systems, and (2) major structural discontinuities (e.g., anticlines) located in areas with potential for increased recharge are no flow boundaries for regional systems. It is important to recognize and identify the basis for the various aspects of the conceptual model. Those aspects of the conceptual model based on general principles must be clearly identified so the data acquisition activities can be designed to gather the information to support or refute them and hopefully convert them into deductive statements whose premises are supported by data and observations. The process of conceptual model development is iterative, involving additional data gathering and refinement. It is important that the decisions made during the course of the iterative process be reviewed for correctness in the light of the new data. During the initial and even in the later development stages, more than one conceptual model may be appropriate. It is important to realize that, in general, it may be easier to disprove a possible conceptual model than to gather the all-inclusive data set to completely validate one. With this in mind, some of the field experiments and data-gathering activities should attempt to gather either or both kinds of information (i.e., information to refute one possible conceptual model and support information for the other). The task of identification of data or experiments to convincingly refute a particular conceptual model must be as carefully undertaken as the task to gain supportive information.

The importance of the conceptual model to the performance assessment process makes one wonder if we should investigate the feasibility of developing a rigorous means for documenting the development of our conceptual models (the assumptions, decisions, etc.) and methods for applying some kind of statistical hypothesis or inference testing methods to quantitatively support the numerous decisions required in developing the final conceptual model or models.

F.3 Performance assessment approach. The performance assessment approach (e.g., detailed models or bounding analysis models) affects how we actually use the field data and the planning of the data acquisition program. If a detailed geochemical submodel was going to be incorporated into a geosphere transport code then a much different data acquisition program would have to be developed than the one developed to support the use of a less detailed sub-model. Data acquisition and testing to support the conceptual model of the subsystem will still have to be carried out to some limited degree, even if a bounding analysis approach is to be used. After all, the correct bounding analysis for the wrong conceptual model is of little use. In the United States, selection of the performance assessment approach, prioritization of tests, and guidance as to the kind and degree of testing will be completed through a process of "performance allocation" [1]. The process of performance allocation will specify which barriers are to be relied upon, the level of performance expected, and the level of confidence expected. This in turn will guide the selection of the performance assessment approach and provide focus for site characterization and experimental activities and eventually provide a means for limiting the number of tests. If no reliance or credit is taken for the site geochemical barrier, for example, then maybe only a limited geochemical data acquisition program aimed at determining boundary conditions for the waste package model is all that is required.

F.4 Sampling size and frequency of sampling; instruments used to gather data; methods used to interpret data. Sampling size, frequency of sampling (both space and time), kinds of tools or instruments, and the methods used to interpret and extrapolate data are too integrally related to discuss separately, but individually they are no less important. In fact, it should be realized that one of the biggest problems associated with modeling subsurface systems is one of characterization and the associated question of dealing with the spatial variability present in these systems. Observations of response and measurements of parameters can only be made at "points" within the system. In reality these "points" are the averaged value for the response or parameter over a specific sample size, determined by the measuring instrument at a mean location in space and time. Characterization of the variability of these parameters in space and time, however, is typically required to model and make performance assessment predictions. The "point" information needs to be extrapolated over the spatial and time domains, and must provide an adequate source for some degree of quantification of spatial and temporal variability in parameters and observations (spatial correlation lengths, directionality). **How much of this variability (i.e., which components of the spatial and temporal frequency domains) must we capture for our purposes?** The uncertainties related to spatial variability of observations and parameters poses only one aspect of the total problem. When we only deal with quantities that can be directly measured like temperature, density, and pressure, this aspect of extrapolation is the only aspect. A more complicated situation arises for parameters (e.g., permeability and dispersivity) that cannot be measured directly, but must be determined indirectly from perturbation-response observations (e.g., pump testing or tracer injection experiments) through inverse modeling techniques.

There are a series of technical issues that arise regarding the relationship between the sampling scale (measurement discretization), sampling frequency, extrapolation and interpretation methods, and the discretization of time and space used in our performance assessment models. A few examples of such issues that need to be considered and addressed when we utilize field data in our models and as we are planning and designing data acquisition programs include the following:

- How should small-sample data be averaged to obtain equivalent large-sample estimates for our performance assessment models? Is it necessary, and is it appropriate?

- What effect does variability in sample size have on our ability to obtain estimates for the spatial distribution of the data set, and what effect will this have on our estimates of spatial correlation lengths?

- How important is it that many of the parameter interpretation theories were developed for a homogeneous world and the real world is heterogeneous?

- For inversely determined parameters in a heterogeneous world, what is the appropriate relationship between:

  - the perturbation stimulus (e.g., pumping rate, screened interval and well diameter, and duration of test)
  - kind, number, locations, and sampling size of response observations (e.g., head or pressure, number of observation wells, diameter, screened interval and distribution around the pumped well)
  - model used for test interpretation
  - the sampling size of the test (e.g., over what volume of rock was the parameter averaged?)
  - the band width of spatial frequencies (i.e., spatial correlation lengths) the test can detect.

The potential importance of these parameter identification and extrapolation steps on the predicted results requires that these kinds of questions be identified and investigated. It is important to determine how much detail we need and at what sampling scale we need to measure. Much of the currently available or developing performance assessment methodology and continuing research and development efforts are related to quantifying or bounding the uncertainty associated with these issues. The utility of existing and developing uncertainty analysis methods combined with sensitivity analysis techniques can greatly improve our programs aimed at gathering new data to further reduce the uncertainty in the characterization of our system. Existing and developing stochastic modeling approaches and methods [6,7], statistical inverse methods [8] (which in some respects are automatic model calibration techniques), and the growing use of geostatistical methods such as kriging is related to the recognized importance of spatial variability in parameters and the difficulty this creates with characterizing the site and developing the conceptual model. Synthetic data sets [9] can also provide a means to determine how much detail we need. If we can tell from our samples how much of the spatial variability we don't know, then maybe synthetic studies can help us determine what we need to find out.

G. Summary Description of the Link Between Laboratory/Field Observations and Models

The linkage between model development and field/laboratory observations might be best described as the iterative and hierarchical program of site and system characterization for development of an observational and confirmatory data base necessary to develop, improve, and support our conceptual models for site and system behavior. Generating confidence that our performance assessments are reliable, realistic, and can provide the correct information for judging the efficacy of disposal is a process that must be pursued in intimate association with site characterization and laboratory and field experiments. The data-acquisition programs will consist of data gathering and experiments to demonstrate understanding at various spatial and time scales and degrees of complexity (more interacting phenomena). The experiments at the various scales will necessarily involve both well-characterized (laboratory) and incompletely characterized (field) experiments. Accounting for the decreasing characterization certainty that accompanies the increasing spatial and time scales, and understanding the implications of the decreasing certainty, will be an important aspect of the link between models and observations. The means for determining when sufficient characterization has been achieved will be through the performance allocation process (goals and confidence levels), and by understanding the implications of characterization uncertainty, along with methods for bounding and/or quantifying this uncertainty. During each iteration in the data acquisition program performance, allocation goals and system understanding are reviewed, and performance goals and confidence levels are revised if necessary. Current site and system characterization data, conceptual models, computer models, and any other appropriate performance assessment tools and approaches are used to design additional needed confirmatory testing or data-gathering programs to improve characterization and understanding in order to meet current performance goals within target confidence levels.

3. REFERENCES

1. J. A. Lieberman, S. N. Davis, D. R. F. Harleman et al. 1985. "Performance Assessment National Review Group." Weston Report RFW-CRWM-85-01, Rockville, Maryland.

2. U.S. Department of Energy. 1985. Part 1, "Overview and Current Program Plans"; Part 2, "Information Required by the Nuclear Waste Policy Act of 1982." Volume 1 of Mission Plan for the Civilian Radioactive Waste Management Program. U.S. Department of Energy, Office of Civilian Radioactive Waste Management, Washington, D.C.

3. G. E. Grisak, J. F. Pickens, J. D. Avis, and D. W. Belanger. 1985. "Principles of Hydrogeologic Investigations at Depth in Crystalline Rock." In Hydrogeology of Rocks of Low Permeability, Part 2 (Proceedings), pp. 52-71. International Association of Hydrogeologists, Tucson, Arizona.

4. C. S. Simmons and C. R. Cole. 1985. "Guideline Approach." Volume 1 of Guidelines for Selecting Codes for Ground-Water Transport Modeling of the Low-Level Waste Burial Sites. PNL-4980, Pacific Northwest Laboratory, Richland, Washington.

5. J. Bear. 1985. "Conceptual and Mathematical Modeling of Groundwater Flow and Pollution: An Overview." Paper presented at the American Society of Civil Engineers Hydraulics Specialty Conference, August 12-17, 1985, Lake Buena Vista, Florida.

6. P. M. Clifton, B. Sagar, and R. G. Baca. 1985. "Stochastic Groundwater Traveltime Modeling Using a Monte Carlo Technique." In Hydrology of Rocks of Low Permeability, Part 1 (Proceedings), pp. 319-331. International Association of Hydrogeologists, Tucson, Arizona.

7. B. Sagar and P. M. Clifton. 1985. "Stochastic Groundwater Flow Modeling Using the Second-Order Method." In Hydrogeology of Rocks of Low Permeability, Part 1 (Proceedings), pp. 498-512. International Association of Hydrogeologists, Tucson, Arizona.

8. E. A. Jacobsen. 1985. "A Statistical Parameter Estimation Method Using Singular Value Decomposition with Application to Avra Valley Aquifer in Southern Arizona." Ph.D. Dissertation, The University of Arizona.

9. C. R. Cole, H. P. Foote, D. A. Zimmerman, C. S. Simmons. 1985. "Understanding, Testing, and Development of Stochastic Approaches to Hydrologic Flow and Transport through the Use of the Multigrid Method and Synthetic Data Sets." Paper presented at the International Symposium on the Stochastic Approach to Subsurface Flow, 3-7 June 1985, Fontainbleau, France.

PROSPECTS OF MODEL VALIDATION AGAINST FIELD/LABORATORY OBSERVATIONS

K. Andersson
Swedish Nuclear Power Inspectorate

ABSTRACT

Code verification, validation, sensitivity and uncertainty analysis are discussed as elements of the achievement of confidence in models for performance assessments of final repositories for nuclear waste. The INTRACOIN- and HYDROCOIN projects are described as two international projects dealing with these issues. Some problems related to model validation are illustrated with results from INTRACOIN and the OECD/NEA Stripa Project and natural analogues are also briefly dealt with in this context. Furthermore, current trends in the development of geosphere models are discussed with the focus on validation. Considering experimental difficulties in model validation, new experimental information, present status of models and the time schedules of nuclear waste management programmes, the need for comprehensive efforts for validation is emphasized. It is stressed that these efforts will require an efficient cooperation between laboratory and field experimentalists, experts in natural analogues, geologists, model developers and users.

POSSIBILITES DE VALIDATION DES MODELES EN FONCTION DES OBSERVATIONS SUR LE TERRAIN/EN LABORATOIRE

RESUME

La vérification, la validation, ainsi que l'analyse de sensibilité et d'incertitude, des programmes de calcul sont examinées en tant que facteurs contribuant à raffermir la confiance à l'égard des modèles destinés à évaluer les performances des installations d'évacuation définitive de déchets nucléaires. L'auteur décrit deux projets internationaux - INTRACOIN et HYDROCOIN - qui traitent de ces questions. Il présente les résultats issus du projet INTRACOIN et du projet OCDE/AEN de Stripa pour illustrer certains problèmes liés à la validation des modèles et évoque aussi brièvement, dans ce contexte, certains phénomènes analogues se produisant dans la nature. En outre, les tendances actuelles dans l'élaboration des modèles de la géosphère sont analysées sous l'angle de la validation. Vu les difficultés expérimentales suscitées par la validation des modèles, les nouvelles données expérimentales disponibles, l'état d'avancement des modèles et les calendriers des programmes de gestion des déchets nucléaires, l'auteur insiste sur la nécessité de consacrer des efforts d'ensemble à la validation des modèles. Il met l'accent sur le fait que ces efforts exigeront une coopération efficace entre les expérimentateurs travaillant en laboratoire et sur le terrain, les spécialistes des phénomènes analogues se produisant dans la nature, les géologues, ainsi que les concepteurs et les utilisateurs des modèles.

1. INTRODUCTION

The ultimate aim of development of performance assessment models is the use of these models in safety evaluations of proposed final repositories or in feasibility studies of disposal concepts. These evaluations and studies are accomplished both by programme responsible organisations as well as regulatory bodies, in the latter case most commonly in a review stage. Performance assessment models also have an important rôle in connection with making priorities of research and development programmes because by carrying through performance assessments, important uncertainties or lack of knowledge can be identified and eventually quantified. The optimization of the design of repositories gives the performance assessment model system a potential third rôle.

The time frames for the realisation of final disposal of spent nuclear fuel and high-level waste varies between different countries. In some countries only general plans exist at the moment, in other countries feasibility studies have been performed without specifying a site or a detailed design whereas other programmes have more explicitly defined goals. The general trend is, however, that disposal studies now leave a generic stage and gradually enter a stage of application in selection of sites and detailed design. This can be illustrated by the Swedish programme as given by the Swedish Nuclear Fuel and Waste Management Co. (SKB) in [1], Figure 1. Following this programme a selection of a few sites for detailed investigations will be done about 1990. A license application based on one site will according to the SKB plan be submitted to the Swedish Nuclear Power Inspectorate (SKI) about the year 2000. For practical reasons it is reasonable to believe that the site characterization prodecure will be concentrated on one particular site in the mid 1990ties.

Thus, in the Swedish example important decisions in e.g. the site selection procedure are planned to be made within 5-10 years. The situtation is similar also for some other programmes (the Swiss programme is one example according to the recently presented Projekt Gewähr [2]). For these decisions and for the forthcoming licensing procedures it is essential that the programme responsible organisations as well as the safety authorities will be able to reach a satisfactory level of confidence in the models used in the performance assessments. This confidence must be based on a firm conviction that the models properly describe the actual processes occurring in a repository and in its surroundings. It is also necessary to have a comprehensive understanding of the characteristics of the mathematical models and their limitations as well as a thorough knowledge of the conceptual and numerical uncertainties involved in the performance assessments.

It should be emphasized that the demands on the performance assessment models as predictive tools increase when the disposal programmes leave the generic stage with feasibility studies and enter a stage of application in site selection, detailed design and optimization. Also this fact can be illustrated with the Swedish example. In June 1984 the Swedish Government granted applications to load the two reactors Forsmark 3 and Oskarshamn 3. This decision which was based on an extensive review process including Swedish as well as foreign organisations (see e.g. [3] and [4]) means, that the method for final disposal presented in the KBS-3-report has been considered acceptable with regard to safety and radiation protection. During the review process, however, a number of issues were pointed out by the review organisations which will have to be thoroughly investigated in the SKB research and development programme before a repository can be licensed. Some of these issues especially concern the model description of transport of groundwater and radionuclides in fractured rock and the validation of these models (see e.g. [5] and [6]). Similar conclusions are made in the Swiss work Projekt Gewähr [2].

The performance assessment model system as applied so far can for practical reasons be divided into four model groups: near field models, groundwater flow models, models for nuclide transport in the geosphere and biosphere models for transport and dilution. These four model groups are coupled as shown in Figure 2. Performance assessments can be made with two complementary approaches: using detailed deterministic models for each of the four groups or with an integrated model system for probabilistic calculations using comparatively simple submodels.

This paper deals with certain aspects of geosphere (groundwater and radionuclide) modelling with the focus on model validation. The survey is concentrated on the description of transport processes in crystalline rock, but it is believed that the general conclusions to a great extent are valid also for other types of host media for final repositories.

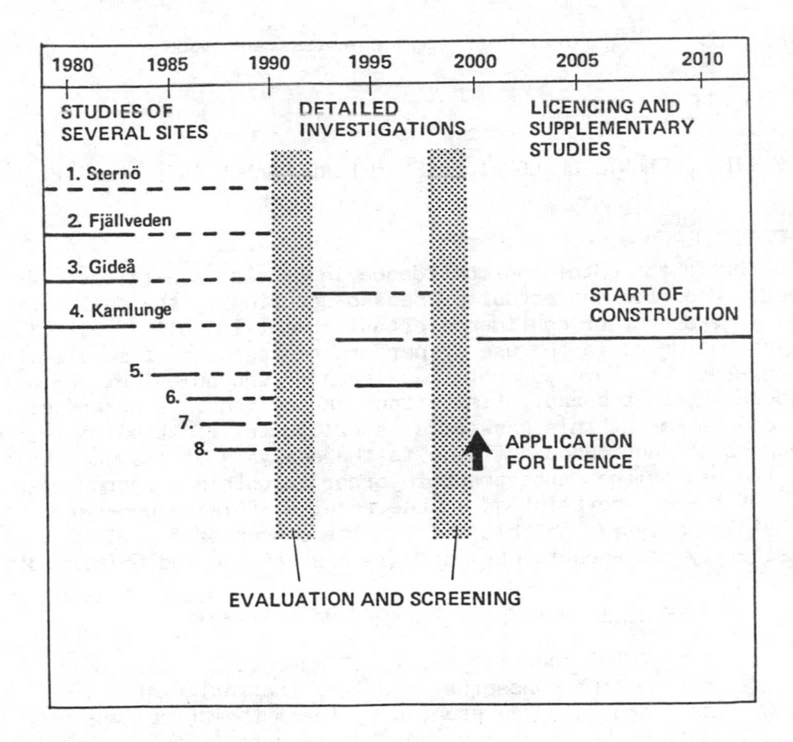

Figure 1. Time-schedule for site-selection studies in the Swedish programme according to the SKB plan (SKB-TR 85-01)

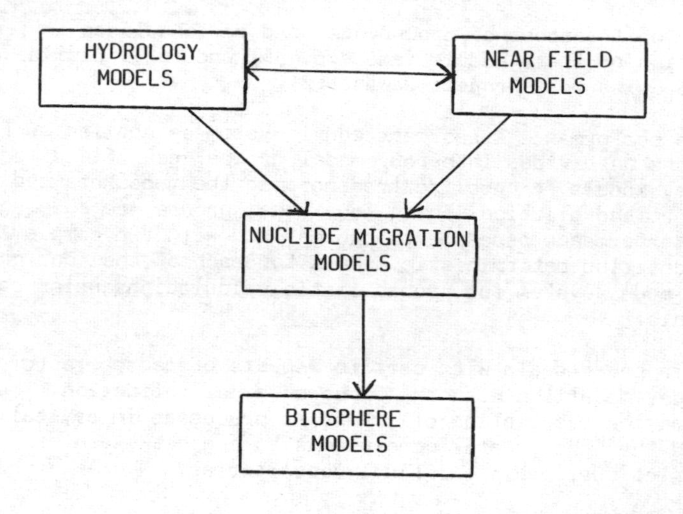

Figure 2. Performance assessment models

2. THE ELEMENTS OF CONFIDENCE IN PERFORMANCE ASSESSMENT MODELS

A. Concepts

The basis for obtaining confidence in models is a firm conviction that they properly describe the actual processes occuring in the real system to be described. This aspect of confidence establishment is called validation. In order to have confidence in the use of performance assessment models it is furthermore necessary to have adequate knowledge of the numerical characteristics of the models, their accuracy, limitations in the range of parameter values, etc. The achievement of this knowledge is called verification. A third element in the process of confidence building is the analysis of the uncertainties involved in the performance assessment in order to obtain a thorough knowledge of the impact of these uncertainties on the results. This is done with sensitivity and uncertainty analyses. In this section the concepts of verification, validation, sensitivity and uncertainty analysis are defined and briefly discussed.

In [7] code verification is defined as follows:

> A computer code is "verified" when it is confirmed that the conceptual model of the real system is adequately represented by the mathematical solution. Verification can thus be carried out, for example, by intercomparison of codes and by comparison of numerical codes with analytical solutions.

Following this definition code verification can be a simple and straight-forward task. Verification exercises can however become rather extensive if they include comparisons between the efficiency of different solution algorithms, comparisons between different dicretisation strategies, convergence tests etc., especially in three dimensions.

The concept of validation is also defined in [7]:

> A conceptual model and the computer code derived from it are "validated" when it is confirmed that the conceptual model and the derived computer code provide a good representation of the actual processes occurring in the real system. Validation is thus carried out by comparison of calculations with field observations and experimental measurements.

In this definition it is foreseen that for model validation information from laboratory or field experimental measurements as well as from field observations may be needed. In the context of waste disposal it is clear that "full validation" of performance assessment models in the meaning of complete confirmation of used theories and parameter values can never be achieved. It is more a matter of a process to gain confidence in the models with the aim of achieving reasonable assurance that they give a good representation of real processes. This can be done by using information from well defined laboratory and field experiments in short time scales and from natural analogues which give information in long time scales but usually with less well defined initial and boundary conditions.

Due to the difficulties involved in model validation and other uncertainties in specifying the repository systems, performance assessments will always suffer from uncertainties. These uncertainties can either be of conceptual nature or be quantifiable within a certain conceptual model. For obtaining confidence in model application it is essential that uncertainties can be handled in an appropriate way using sensitivity and uncertainty analyses with deterministic or probabilistic tools.

3. THE INTRACOIN AND HYDROCOIN PROJECTS

In Section 2 verification and validation as well as sensitivity and uncertainty analysis were described as complementary elements in the process of gaining confidence in performance assessment models. Since 1981 two international collaboration projects dealing with these aspects of geosphere modelling, INTRACOIN and HYDROCOIN have been initiated and managed by SKI. The project INTRACOIN, which dealt with nuclide migration codes began in 1981 and is now in its final stage. The HYDROCOIN project dealing with groundwater flow codes was started in 1984 and is scheduled to continue until 1987.

The overall objective of the two projects have been to obtain improved knowledge of the influence of various strategies for geosphere modelling on the safety assessment of final repositories for nuclear waste. To this end calculations have been made with different mathematical models used by a number of organisations. The studies comprise:

a) the impact on the transport calculations using different solution algorithms,

b) the capabilities of different models to describe field measurements, and

c) the impact on the transport calculations of incorporating various physico-chemical phenomena.

The projects have thus been performed at three levels corresponding to verification (level one) validation (level two) and sensitivity and uncertainty analysis (level three). Apart from level two of INTRACOIN, which will be described later in this paper, the general results from INTRACOIN and the present status of HYDROCOIN are briefly described here.

Level one of INTRACOIN dealt with the verification of 22 computer codes handling the transport equation. The complexity of the participating codes varied from one-dimensional advection-dispersion models with linear sorption to models including e.g. matrix diffusion or equations for hydrology and heat transport. The verification exercise was executed with seven different cases ranging from simple one-dimensional transport in a porous medium with constant parameters to two dimensional transport in a fractured medium with diffusion into the rock matrix. In some cases analytical solutions were used for the comparisons, in other cases intercomparisons of results from different solution algorithms were performed. The results have been reported in [8]. In general the results show good agreement between the different codes. Most of the deviating results have been explained, either by differences in the implementation of boundary conditions, deviating interpretation of input data or as a result of truncation errors.

At the third level of INTRACOIN a central case with 12 variations were defined for an uncertainty analysis. The central case can be described as advection-dispersion with a single nuclide (neptunium-237) in a fractured medium. Although very limited in its scope the INTRACOIN level three illustrates the importance for the performance assessment of validating and quantifying sorption, matrix diffusion and dispersion. The results will be presented in a final report for INTRACOIN levels two and three [9]. Some of the participants have also published reports of their work, see e.g. [10].

As concerns HYDROCOIN the work is in progress. The progress of the study is briefly reported in a series of Progress Reports [11]. The bulk of calculations on level one has been made on seven cases, illustrated in [11]. The results show good agreement with respect to the scalar functions pressure and temperature but do also indicate that difficulties in some cases can arise in the calculation of velocities and pathlines. Although two cases, one based on a heater experiment in a Cornish quarry and one with unsaturated flow in consolidated tuff using data from the Nevada test site, have been defined for level two, it remains to set up the main cases based on field experiments for model comparisons. The level three of HYDROCOIN has not yet been defined but it is foreseen that rather extensive sensitivity and uncertainty analyses will be accomplished.

4. SOME PROBLEMS RELATED TO THE VALIDATION OF GEOSPHERE MODELS

In this Section the conceptual geosphere models used in performance assessments related to crystalline rock are briefly described. Furthermore, some problems with the validation of these models are illustrated with results from level two of INTRACOIN and some experiences from the Stripa project. Finally, the potential use of natural analogues in the context of model validation is shortly discussed with reference to a recent assessment.

A. Conceptual models used in performance assessments

The modelling of groundwater flow in extensive performance assessments like KBS-3 [12] and Projekt Gewähr [13] has been made with the underlying assumption that the rock can be described as a porous homogeneous medium, i.e. using Darcy's law. This assumption requires that the rock properties can be described by averaging over a region that must be large in relation to the geometrical properties of the fracture system. The scale at which this can be done depends on fracture length, frequency, permeability etc. The hydrology calculations made e.g. in KBS-3, using information on the extent and properties of different hydraulic units and groundwater levels, were performed applying numerical approximations in three dimensions, resulted in a head distribution within the regarded region from which flow rates and flow paths could be derived.

The radionuclide transport calculations use information from the hydrology calculations in the form of flow rates and flow paths. The models typically incorporate dispersion, sorption and eventually matrix diffusion into the transport equation (usually in one dimension). Applying initial conditions deemed as realistic (or pessimistic) and physically reasonable boundary conditions, the calculations result in rates of inflow to the biosphere of relevant nuclides. In addition to the assumptions inherent in the hydrology, the description of radionuclide movement use a number of conceptual assumptions regarding dispersion and retardation. It is e.g. assumed that a number of different dispersion effects can be described with a single dispersion term in the transport equation. Furthermore, the retardation is either described as sorption on fracture surfaces using the K_d-concept or as diffusion into the rock matrix combined with sorption on the inner surfaces of the matrix.

B. The INTRACOIN experience

At the level two of INTRACOIN three cases were set up [9]:

a) a synthetic two-dimensional case describing a rectangular cut through a system with two media differing in physical properties,

b) a case simulating a dual-tracer injection-withdrawal test in a sandy aquifer in Canada [14],

c) a fractured medium case based on a two-well tracer experiment near the lake Finnsjoen in Sweden [15].

The aim of cases b) and c) was to examine the capabilities of the nuclide migration codes to describe in-situ measurements. The selection of these two field experiments was carefully made with assistance from two groups of experts on porous and fractured media respectively. The experiments were judged to be two of the best available experiments for the purpose of comparison with model calculations. Since this paper deals primarily with the fractured type of geologic media, case c) is described here with some results. Detailed information is available in [9].

The layout of the tracer experiment is shown in Figure 3. Tracers were injected in one borehole between two packers in a high permeability zone at a depth of about 100 m. In another hole 30 m distant, water was continuously pumped out creating a drawdown zone into which the tracers from the injection hole would flow. The tracers used in the INTRACOIN calculations were iodide and strontium. For the calculations the field experimental data were complemented with laboratory measurements of sorption equilibria and diffusivity data on granite from the Finnsjoen area.

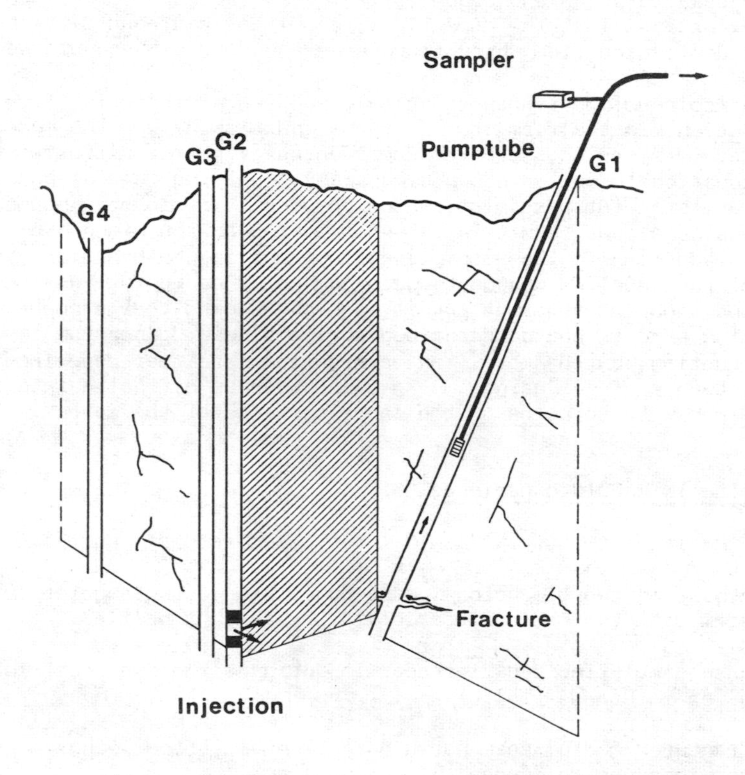

Figure 3. Cutaway view of test site and instrument set-up for the field experiment used for INTRACOIN level two case for fractured media

One conclusion from the comparisons between the experimental data and the calculational results is that existing nuclide migration codes can reproduce results from field experiments. However, it is evident that the available data leave too many degrees of freedom for parameter fitting in the calculations to allow a model validation in a stringent sense. Thus it was possible to simulate the experimental data with several different interpretations of the physical situation (different mechanisms for dispersion, diffusion of the tracers into fracture filling material, rock matrix or volumes of stagnant water) although the best fits were obtained with models that include matrix diffusion. Moreover, results from the parameter fitting also show deviations from laboratory data on the diffusivity of iodide and strontium into the rock matrix.

One problem with the Finnsjoen experiment from the validation point of view is the uncertainty with respect to the heterogeneities and flow channels in the rock between the two boreholes. Another problem identified in the INTRACOIN work is that the available laboratory data had not been obtained on samples from the field test site.

One obvious conclusion from INTRACOIN level two is that there is a need for more carefully designed experiments for the purpose of model validation. To achieve this a close collaboration between field and laboratory experimentalists, geologists and modellers is necessary.

C. Some results from the Stripa Project

Some of the conclusions from INTRACOIN level two are supported with results from the OECD/NEA Stripa project managed by the Swedish Nuclear Fuel and Waste Management Company, SKB. In [16] results from investigations in three fractures located in drifts at a depth of 360 m i the Stripa mine in Sweden are reported. The fractures which are clearly visible in the drifts have a natural flow of water. Hydraulic tests and migration tests with nonsorbing as well as sorbing tracers were performed. The injections took place at 5-10 m distance from the roof of the drifts.

The results from runs with nonsorbing tracers and data on sorption and porosity obtained in the laboratory were used to predict the breakthrough curve for a sorbing tracer. The predicted breakthrough curve was later compared with the experimentally obtained breakthrough curve.

Three different models were used to analyse the experimental data:

1. an advection-dispersion-surface sorption model,

2. an advection-dispersion-matrix diffusion model,

3. an advection-channelling-matrix diffusion model.

The effects of channelling, matrix diffusion and diffusion into stagnant zones of water could not be distinguished from the analysis of breakthrough curves only. In the Stripa investigation matrix diffusion was however observed by direct observation of penetration depths and channelling was observed by using a large amount of collecting points. The presence of stagnant "lakes" was not observed directly but inferred from general observations on the necessity of points of contact between rock blocks.

The results indicate that field experiments, possibly also in a larger scale, can give very useful results for the validation of nuclide transport models if they are especially designed for this purpose. Information available from the Stripa project indicate that the concepts for the modelling of fracture flow should be reviewed. Thus the authors of [16] say:

> Our present concept of flow in a fracture is that there are different channels which may mix their waters at irregular distances. The extent of the channels and the mixing distances are unknown at present. The channels may have zones of stagnant or near stagnant water which may be reached mainly by diffusion. The matrix is porous and contains stagnant water in the micropores of the matrix which also are reached by diffusion only. Fissure coating or filling material may vary in composition and thickness as well as in porosity along the channels and influence the tracer transport. There are models which account for all these effects but there is no model which takes them into account at the same time.

In [17] results from Stripa are reported that give strong indications for "channelling".

D. Natural Analogues

It is believed that information from natural analogues should be useful in the process of validating geosphere models, especially with respect to the long time scales which can be studied. In [18] an extensive compilation and evaluation of different natural analogues with respect to the assessment of deep disposal is made. Information in this subsection should be regarded mainly as extracts from this report. Regarding analogue studies of radionuclide retardation processes three requirements are set up in the report:

- a well defined source input function,
- a well defined past hydrogeological environment,
- a resultant measurable concentration profile.

These requirements are found hard to meet sufficiently in detail in order to allow e.g. realistic back-calculation of retardation data. So far the information gained from natural analogues have been more of a qualitative nature. One of the difficulties encountered in the study of elemental profiles along flow paths around ore bodies is the difficulty to separate control by release as opposed to retardation processes.

Concerning matrix diffusion the authors of [18] believe that it is reasonable to conclude that laboratory-oriented studies indicate that the effect takes place. They conclude however that at present natural analogue studies have failed to provide conclusive evidence for the process or its likely magnitude in crystalline rock. The major difficulty is the suitability of samples investigated to date due to possible near-surface mechanical and chemical effects.

The review made in [18] highlights the lack of good analogues in the time range of 1 000 to 100 000 years, that is between archeological analogues and the bulk of geological analogues, where timescales of 10^6-10^7 years are the most commonly considered. It is furthermore pointed out in [18] that almost all the useful information which can be obtained from natural analogues relates to chemical processes. Groundwater flow modelling does not seem amenable to analogue study.

In spite of the various difficulties involved in the use of natural analogues for model validation, it is concluded in [18] that analogues have an undoubted potential in answering some of the important questions regarding the very long term predictability of processes in the natural environment.

Furthermore, a number of studies with potential possibility to enhance our understanding of major processes are suggested. It is believed though, that the rôle of analogues will be in verification of specific processes of importance which can be decoupled for reliable observation.

5. CURRENT TRENDS IN MODEL DEVELOPMENT

The conciousness about certain basic problems inherent in existing performance assessment model systems, new experimental information and the fact that some effects and couplings between effects with potential importance have not yet been taken into account are all stimulating factors for model development. Present developments can be divided into three main topics relevant for model validation:

1. Development of new and refined models for the detailed description of the basic processes involved in the transport of groundwater and radionuclides.

2. Incorporation into the models of effects that have not been taken into account so far and the couplings between these effects (i.e. hydrological, thermal, mechanical and chemical effects).

3. Development of integrated model systems for performance assessment using e.g. probabilistic techniques.

A. New and refined basic models

The developments in the first group that will be discussed here are stochastic fracture models, models describing channelling, matrix diffusion models, geochemical models and three-dimensional nuclide transport models.

The question of the scale of the validity of groundwater flow models describing crystalline rock as a porous homogeneous medium is of fundamental importance for the performance assessment. During the latest years development of models describing the rock as a fracture system with stochastic distribution of rock properties such as length, frequency, direction and permeability has begun. These models have so far been used in two-dimensional applications illustrating certain aspects of the problem at hand, see e.g. [19] and [20]. It is believed that the applicability of this kind of models in the performance assessment can be better understood within a few years.

As mentioned in Section 4 there are strong indications that the main part of the water flow takes place not just in discrete fractures but also in discrete channels within these fractures. This channelling effect has a potential for increasing the dispersion with the possibility for "fast channels" to the biosphere and for decreasing the fraction of fracture surfaces which are available for sorption thus decreasing the retardation. Projections have been made indicating a very strong influence of channelling on the migration of radionuclides [21]. If channelling proves to be an important effect this can have a strong impact on both hydrology and migration modelling.

Models including matrix diffusion have already been used in performance assessments [22], [23] and can thus not been regarded as an entirely new concept. The magnitude of this effect has, however, been discussed. This has led to the concept of "limited matrix diffusion" in the Swiss example [24].

In radionuclide transport models the sorption of radionuclides on fracture surfaces is usually described with the K_d concept. This approach cannot deal with expected variations in sorption properties of water-rock systems due to chemical factors (pH, redox conditions etc.) and variations in surface properties of sorbing phases [25]. A number of computer models for calculations on groundwater-rock reaction chemistry exist. A brief description of the abilities of such models can be found e.g. in [26]. The geochemical models have a large potential impact on the nuclide transport calculations either by giving more detailed input data or by being directly coupled to the transport models in an integrated system. For these applications further mathematical development is needed. Furthermore, more comprehensive data bases must be developed.

Another line of development in nuclide transport modelling is the extension from one- and two-dimensional calculations to three dimensions. The circumstance that the applications become more and more site-specific will most probably increase the interest in three-dimensional nuclide transport calculations. The fact that groundwater transport modelling already to a large extent is made in three dimensions should stimulate this development. Some three-dimensional nuclide transport codes exist today. The SWIFT code [27] is one example of a code which combines the groundwater and nuclide transport. The NAMSOL code is another example of a three-dimensional nuclide transport code 28 .

B. Coupled effects

In order to achieve confidence in geosphere modelling it is necessary to understand the combined effects of many different processes that may affect the groundwater and radionuclide transport. The major processes involved are hydrological, thermal, mechanical and chemical.

Among the four categories of processes there can be eleven types of couplings of various importance. In [26] different coupled effects are identified and discussed in a systematic manner. In order to understand the importance of the various effects and to be able to rank them in specific repository environments a large effort of a multidisciplinary nature seems necessary.

One example of a coupled effect which already has been investigated to some extent is the thermally driven groundwater flow, see e.g. [29]. One extensive validation exercise with a coupled water and heat flow code has been carried out by Lawrence Berkeley Laboratory against a series of field experiments with injection and subsequent production of hot water into a confined aquifer [30]. For HYDROCOIN level two an example using data from a heater experiment in Cornwell have been defined.

The application in performance assessments of other coupled processes may be somewhat more remote although work on e.g. the coupled thermal-hydraulic-mechanical phenomena already have been made [31]. Mechanical effects on the flow properties of the rock can be created either directly by the excavation of a repository or indirectly via thermal effects on the rock apertures. In [26] a detailed analysis of these and other mechanical processes is recommended as the conductivity is highly sensitive to the change in aperture of existing fractures and to the generation of new fractures.

As said in Section 4 progress in the study of geochemical effects on nuclide migration is highly desirable. Furthermore, several coupled chemical processes in the near field which are not yet thoroughly investigated are discussed in [26].

C. Probabilistic techniques

During the latest years probabilistic integrated performance assessment models have been developed, see e.g. [32] and [33]. In a probabilistic analysis several submodels, typically representing the near field, the geosphere and the biosphere are run in sequence a number of times with a probabilistic choice of parameter values. The result is a probability density function of consequence estimates. This type of analysis put certain restrictions on the complexity of submodels due to feasibility with respect to computer time.

It is believed that probabilistic performance assessment models should be useful tools complementing detailed deterministic models. Uncertainty and sensitivity analyses using both kinds of methology should be of great value not just in safety assessments but also as instruments for the direction of research into high-priority areas. In cases of uncertainties concerning the conceptual models it may be appropriate to execute parallel probabilistic calculations using different concepts in the submodels.

The validation of probabilistic model systems may be divided into three parts: validation of submodels, validation of probability distributions of parameter values and validation of the statistical procedures. The validation of submodels can be done as verification against detailed deterministic models. The validation of the probabilistic methods is an exercise of a mathematical-statistical nature although very important. The validation of the probability distributions may be the most difficult one which has to be done using sets of laboratory and field data for the parameters used in the submodels.

6. DISCUSSION

Some uncertainty and sensitivity analyses accomplished e.g. in the SKI review of KBS-3 [34], in the Swiss Projekt Gewähr [2] and in the INTRACOIN project [9] have shown that reasonable variations in conceptual assumptions and parameter values can give differences of many orders of magnitude in the output from geosphere models for important nuclides. Furthermore, the results from the extensive hydrology modelling efforts in both the Swedish and Swiss examples have been very little used for the nuclide transport calculations. As compared to the results from the hydrology calculations very conservative values have been used for flow rates and migration distances (i.e. the coupling between the two model groups has been weak).

These experiences reflect the circumstance that it is presently in some cases difficult not just to quantify parameters but also to decide which conceptual models are appropriate. This fact is in part due to difficulties involved in the validation of geosphere models. Laboratory experiments have certain drawbacks in this respect:

- the sampling procedure can change the properties of the samples due to e.g. stress release,

- the time scales for the experiments are very short compared to the time scales involved in performance assessments,

- there are problems to simulate the water properties at repository depth.

The use of field field tracer experiments for validation exercises of migration models have so far also encountered a number of difficulties due to e.g.:

- uncertainties concerning the flow situation,

- lack of laboratory data from the field test sites.

Moreover different conceptual models tend to give similar results with respect to break-through curves, a fact which increases the difficulties to separate different effects in comparisons between model predictions and experimental data.

Natural analogues have an obvious advantage in comparison with experiments with respect to time scale. Also the use of analogues have, however, a number of drawbacks which have to be considered:

- problems in defining initial conditions,

- problems in defining the hydrological situation,

- problems in the sampling procedure.

It can thus be concluded the three considered methods for model validation all have their specific problems. It should, however, be possible to do substantial progress with each of the methods by designing experiments directly for the purpose of validation and by investigating natural analogues systematically considering their specific rôle in the validation process. Moreover, an integrated effort to combine information from laboratory and field experiments as well as from natural analogue studies should be of great value in this process.

Valuable validation exercises concerning coupled processes are in some cases quite feasible to perfrom [30] but may in other cases they require rather extensive efforts. It is e.g. suggested in [26] that large-scale experiments should be performed to investigate thermomechanical effects on groundwater flow.

This paper deals with issues concerning validation of performance assessment models. It is important to keep in mind that the validation of underlying assumptions in the application of these models concerning e.g. redox conditions may be equally significant.

7. CONCLUSIONS

The existing model system for performance assessments of final repositories for nuclear waste in crystalline rock has been used in extensive generic studies with the aim of demonstrating the feasability of certain disposal concepts. In the case of the Swedish KBS-3 concept this model system has been regarded sufficient for this purpose.

It has, however, also been expressed by responsible organisations in several countries in e.g. the KBS-3 review that the state of knowledge in several areas must be further developed before detailed site specific analyses to be used in licensing procedures should be performed. For example a number of effects and couplings between different effects of potential importance have not yet been taken into account in overall performance analyses. Furthermore there are some basic conceptual problems involved e.g. with respect to the description of transport processes in fractured media.

Within the next few years a number of models aimed at improving our understanding of these and other related matters will be developed. It seems evident that the appropriate rôle in the performance assessment of different currently existing and forthcoming models will be uncertain for some years. The main reason for this is that existing experimental and natural data leave big uncertainties for model developers and users both with respect to conceptual assumptions and parameter values. Results from the INTRACOIN project illustrate this fact.

The awareness that future model systems for performance assessment must be more firmly based on experimental and natural evidence than current model system put a strong emphasis on model validation. Considering the fact that nuclear waste disposal programmes now gradually leave the generic stage entering applications it is clear that the need for "validated" models becomes more urgent.

Taking into account that the present situation is characterized by a diverging stage of model development and application, high costs and long time frames for experimental (especially large scale) activities as well as experienced difficulties in comparisons between model predictions and experimental results, the validation issue may seem difficult to handle. It should, however, be possible to develop an international concensus on a strategy for the validation work. The objective for such a strategy should be to make it possible to reach concensus about the range of applicability of different modelling approaches and important assumptions in the applications within a reasonable time period (10-15 years).

In developing a strategy for model validation one must be aware of the fact that the full validation in the sense of complete confirmation of model assumptions is impossible, especially with respect to the time scales involved in the performance analysis. The aim of the work is rather to reach reasonable assurance that the models give good representations of the processes occurring in the repository systems.

The needed degree of validation will be different for different model groups and different effects depending on their various rôles in the performance assessment. The adjustment between different efforts should be made within a systems approach and with uncertainty analysis as an important tool. At current stage it is also important to stimulate new model developments and to examine different modelling approaches.

An efficient interaction between model developers and experimentalists is critical for progress. Especially there is a need for further experiments directly designed for the purpose of model validation. The use of natural analogues should be further investigated.

REFERENCES

[1] SKB Technical Report 85-01, Annual Research and Development Report 1984, Swedish Nuclear Fuel and Waste Management Co., Stockholm, June 1985.

[2] Projektbericht, Das Projekt Gewähr (in German), NAGRA, Baden, January 1985.

[3] Review of KBS-3, Plan for Handling and Final Storage of Unreprocessed Spent Nuclear Fuel, Ds I 1984:17, Ministry of Industry, Stockholm, 1984.

[4] Fuelling Licenses for Forsmark 3 and Oskarshamn 3 (in Swedish), Ds I 1984:19, Ministry of Industry, Stockholm, 1984.

[5] Review of Final Storage of Spent Nuclear Fuel - KBS-3 (in Swedish, English translation in print), Swedish Nuclear Power Inspectorate, Stockholm, February 1984.

[6] Comments on Final Storage of Spent Nuclear Fuel - KBS-3, SKI 84:4, Swedish Nuclear Power Inspectorate, Stockholm, February 1984.

[7] Radioactive Waste Management Glossary, IAEA-TECDOC-264, International Atomic Energy Agency, Vienna, 1982.

[8] INTRACOIN, Final Report Level 1, Code Verification, SKI 84:3, Swedish Nuclear Power Inspectorate, Stockholm, September 1984.

[9] INTRACOIN, Final Report Levels 2 and 3, Swedish Nuclear Power Inspectorate, in preparation.

[10] Hodgkinson, D.P., Lever, D.A. and England T.H.
Mathematical Modelling of Radionuclide Transport through Fractured Rock using Numerical Inversion of Laplace Transforms: Application to INTRACOIN Level 3, AERE-R 10986, AERE Harwell, September 1983.

[11] HYDROCOIN Progress Reports, No 1-2, Swedish Nuclear Power Inspectorate.

[12] Carlsson, L., Winberg, A. and Grundfelt, B.
Model Calculations of the Groundwater Flow at Finnsjön, Fjällveden, Gideå and Kamlunge, SKBF/KBS 83-45, Swedish Nuclear Fuel Supply Co./Division KBS, Stockholm, May 1983.

[13] Kimmeier, F., Perrochet, P., Kiraly, L. and Andrews, R.
Simulation des Écoulements Souterrains par Modèle Mathématique entre les Alpes et la Forêt Noire, NTB 84-50, NAGRA, Baden 1985.

[14] Pickens, J.F., Jackson, R.E. and Inch, K.J.
Measurement of Distribution Coefficients using a Radial-Injection Dual-Tracer Test, Water Resources Research, Vol. 17, no. 3, pp 529-544, June 1981.

[15] Gustafsson, E. and Klockars, K.
Studies on Groundwater Transport in Fractured Crystalline Rock under Controlled Conditions using Nonradioactive Tracers, SKBF/KBS 81-07, Swedish Nuclear Fuel and Supply Co./Division KBS, Stockholm, April 1981.

[16] Abelin, H., Neretnieks, I., Tunbrant, S. and Moreno, L.
Final Report of the Migration in a Single Fracture-Experimental Results and Evaluation, Stripa Project 85-03, SKB, May 1985.

[17] Abelin, H., Birgersson, L., Gidlund, J., Moreno, L., Neretnieks, I and Tunbrant, S.
Flow and Tracer Experiment in Crystalline Rocks. Results from Several Swedish in-situ Experiments Presented at MRS Conference on Scientific Basis for Nuclear Waste Management, Stockhom, September 1985.

[18] Chapman, N.A., Mc Kinley, I.G., and Smellie, J.A.T.
The Potential of Natural Analogues in Assessing Systems for Deep Disposal of High-level Radioactive Waste, SKB/KBS 84-16, Swedish Nuclear Fuel and Waste Management Co., Stockholm, August 1984 (NAGRA NTB 84-41).

[19] Long, J.C.S. and Witherspoon, P.A.
The Relationship by the Degree of Interconnection to Permeability in Fracture Networks, Journal of Geophysical Research, Vol. 90, no. B4, pp 3087-3098, March 1985.

[20] Andersson, J.
Predicting Mass Transport in Fractured Rock with the Aid of Geometrical Field Data, Symposium on the Stochastic Approach to Subsurface Flow, Montvillargenne, June 1985.

[21] Neretnieks, I.
A Note on Fracture Flow Dispersion Mechanisms in the Ground Water, Water Resources Research, Vol. 19, no. 2, pp 364-370, April 1983.

[22] Rasmuson, A and Neretnieks, I.
Migration of Radionuclides in Fissured Rock - Results obtained from a Model based on the Concept of Hydrodynamic Dispersion and Matrix Diffusion, SKBF/KBS 82-05, Stockholm, May 1982.

[23] Bengtsson, A., Magnusson, M., Neretnieks, I. and Rasmuson, A.
Model Calculations of the Migration of Radionuclides from a Repository for Spent Nuclear Fuel, SKBF/KBS 83-48, May 1983.

[24] Hadermann, J. and Roesel, F.
Radionuclide Chain Transport in Inhomogenous Crystalline Rocks - Limited Matrix Diffusion and Effective Surface Sorption, EIR-Bericht nr. 551, Swiss Federal Institute for Reactor Research, Würenlingen, February 1985.

[25] Muller, A.B., Neretnieks, I. and Langmuir, D.
Conclusions from an NEA Workshop, Symposium, Proc. 7th Intl. Symp. on the Scientific Basis for Nuclear Waste Mgmt., Boston, November 1983.

[26] Panel Report on Coupled Thermo-Mechanical-Hydro-Chemical Processes Associated with a Nuclear Waste Repository, LBL-18250, Editors: Tsang, C.F. and Mangold, D.C., Lawrence Berkeley Laboratory, July 1984.

[27] Dillon, R.T., Lantz, R.B. and Pahwa, S.B.
Risk Methology for Geologic Disposal of Radioactive Waste: The Sandia Waste Isolation Flow and Transport (SWIFT) Model, SAND-78-1267, 1978.

[28] Atkinson, R., Herbert, A.W., Jackson, C.R. and Robinson, P.C.
NAMSOL User Guide, AERE-R 11365, Harwell Laboratory, May 1985.

[29] Hodgkinson, D.P., Lever, D.A. and Rae, J.
Thermal Aspects of Radioactive Waste Burial in Hard Rock, Progress in Nuclear Energy 11(2), pp 183-218, 1983.

[30] Tsang, C.F. and Doughty, C.
Detailed Validation of a Liquid and Heat Flow Code against Field Performance, LBL-18833, Lawrence Berkeley Laboratory, February 1985.

[31] Noorishad, J., Tsang, C.F. and Witherspoon, P.A.
Coupled Thermal-Hydraulic-Mechamical Phenomena in Saturated Fractured Porous Rocks, Journal of Geophysical Research, Vol. 89, no. B12, pp 10365-10373, November 1984.

[32] Goodwin, B.W.
The SYVAC Approach for Long-Term Environmental Assessments, Presented at the Symposium on Groundwater Flow and Transport Modelling for Performance Assessment of Deep Geologic Disposal of Radioactive Waste & Critical Evaluation of the State of the Art, Albuquerque, May 1985.

[33] Thompson, B.G.J., Duncan, A.G.D. and Hall P.A.
The Development of Radiological Assessment Methods by the Department of the Environment, Proceedings of the Conference on Radioactive Waste Management, B.N.S., London, November 1984.

[34] Andersson, K., Kjellbert, N.A. and Forsberg, B.
Radionuclide Migration Calculations with Respect to the KBS-3-Concept, SKI 84:1, Swedish Nuclear Power Inspectorate, Stockholm, February 1984.

LIST OF PARTICIPANTS LISTE DES PARTICIPANTS

BELGIUM - BELGIQUE

MARIVOET, J., Centre d'Etude de l'Energie Nucléaire (S.C.K./C.E.N.), Boeretang 200, B-2400 Mol

CANADA

ASHBROOK, D.W., Eldorado Resources Limited, 255 Albert Street, Suite 400, Ottawa, Ontario K1P 6A9

JACK, G.C., Atomic Energy Control Board, 270 Albert Street, P.O. Box 1046, Ottawa, Ontario K1P 5S9

JARVIS, R.G., Atomic Energy of Canada Limited, Chalk River Nuclear Laboratories (CRNL), Chalk River, Ontario KOJ 1J0

LYON, R.B. (Chairman), Atomic Energy of Canada Limited, Whiteshell Nuclear Research Establishment, Pinawa, Manitoba ROE 1L0

FINLAND - FINLANDE

PELTONEN, E.K., Industrial Power Company Ltd., Fredrikinkatu 51-53 B, SF-00100 Helsinki

RUOKOLA, E.J., Finnish Centre for Radiation and Nuclear Safety, Department of Nuclear Safety, P.O. Box 268, SF-00101 Helsinki

VUORI, S.J.V., Technical Research Centre of Finland, Nuclear Engineering Laboratory, P.O. Box 169, SF-00181 Helsinki

FRANCE

DURIN, M., Commissariat à l'Energie Atomique, Centre d'Etudes Nucléaires de Saclay, EMT/SMTS/TTMF/Bâtiment 91, F-91191 Gif-sur-Yvette Cédex

LEWI, J., Commissariat à l'Energie Atomique, IPSN/DAS/SASICC/SAED, B.P. n° 6, F-92260 Fontenay-aux-Roses

MANGIN, J.P., Commissariat à l'Energie Atomique, Centre d'Etudes Nucléaires de Fontenay-aux-Roses, DRDD/SESD, B.P. N° 6, F-92265 Fontenay-aux-Roses

FEDERAL REPUBLIC OF GERMANY - REPUBLIQUE FEDERALE D'ALLEMAGNE

HILD, W., Gesellschaft für Strahlen- und Umweltforschung mbH., Institut für Tieflagerung, Abteilung Endlagersicherheit, Theodor-Heuss-Str. 4, D-3300 Braunschweig

STORCK, R., Gesellschaft für Strahlen- und Umweltforschung mbH München, Institut für Tieflagerung, Theodor-Heuss-Strasse 4, D-3300 Braunschweig

ITALY - ITALIE

GERA, F., ISMES S.p.A., Via dei Crociferi, 44, I-00187 Roma

JAPAN - JAPON

MATSUNAGA, T., Department of Environmental Safety Research, Japan Atomic Energy Research Institute, Tokai-mura, Naka-gun, Ibaraki-ken, 319-11

NETHERLANDS - PAYS-BAS

BOSNJAKOVIC, B.F.M., Ministry of Housing, Physical Planning and Environment, P.O. Box 450, 2260 MB Leidschendam

GLASBERGEN, P., RIVM, P.O. Box 150, 2260 AD Leidschendam

KORTHOF,R.M., Ministry of Economic Affairs, P.O. Box 20101, 2500 EC The Hague

PRIJ, J., Netherlands Energy Research Foundation (ECN), Westerduinweg 3, 1755 ZG Petten

SWEDEN - SUEDE

ANDERSSON, K., Swedish Nuclear Power Inspectorate (SKI), Box 27106, S-102 52 Stockholm

BOGE, R., National Institute of Radiation Protection, Box 60204, S-104 01 Stockholm

PAPP, T., (Rapporteur, Session III), Swedish Nuclear Fuel and Waste Management Company (SKB), Box 5864, S-102 48 Stockholm

SWITZERLAND - SUISSE

HADERMANN, J., Swiss Federal Institute for Reactor Research, CH-5303 Würenlingen

HUFSCHMIED, P., Société Coopérative Nationale pour l'Entreposage de Déchets Radioactifs, NAGRA, Parkstrasse 23, CH-5401 Baden

McCOMBIE, C., Société Coopérative Nationale pour l'Entreposage de Déchets Radioactifs, NAGRA, Parkstrasse 23, CH-5401 Baden

VAN DORP, F., Société Coopérative Nationale pour l'Entreposage de Déchets Radioactifs, NAGRA, Parkstrasse 23, CH-5401 Baden

ZURKINDEN, A., Nuclear Safety Department, Federal Office of Energy, CH-5303 Würenlingen

UNITED KINGDOM - ROYAUME-UNI

GRIMWOOD, P.D., Safety and Medical Services, BNFplc, Sellafield Works, Sellafield, Cumbria

JOHNSTON, P.D., (Rapporteur, Session II), Department of the Environment, Romney House, Room A.518, 43 Marsham Street, London, SW1P 3PY

ROBINSON, P.C., Theoretical Physics Division, Bldg. 424 4, AERE, Harwell, Didcot, Oxon OX11 ORA

SMITH, G.M., National Radiological Protection Board, NRPB, Chilton, Didcot, Oxfordshire, OX11 ORQ

UNITED STATES - ETATS-UNIS

COLE, C., Battelle Pacific Northwest Laboratory, Battelle Boulevard, P.O. Box 999, Richland, Washington 99352

CRANWELL, R.M., (Rapporteur, Session IV), Waste Management Systems, Division 6431, Sandia National Laboratories, P.O. Box 5800, Albuquerque, New Mexico 87185

DEFIGH-PRICE, C., (DOE Defense Waste; Environmental Analysis), Rockwell Hanford Operations, P.O. Box 800, Richland, Wa. 99352

RANDALL, J.D., Office of Nuclear Regulatory Research, United States Nuclear Regulatory Commission, Mail Stop 1130SS, Washington, DC 20555

VAN LUIK, A.E., Battelle Pacific Northwest Laboratory, Battelle Boulevard, P.O. Box 999, Richland, Washington 99352

COMMISSION OF THE EUROPEAN COMMUNITIES
COMMISSION DES COMMUNAUTES EUROPEENNES

CADELLI, N., Commission of the European Communities, 200, rue de la Loi, B-1049 Brussels, Belgium

SALTELLI, A., (Rapporteur, Session III), Commission of the European Communities, Joint Research Centre of Ispra, I-21020 Ispra (Varese), Italy

NEA DATA BANK
BANQUE DE DONNEES DE L'AEN

NAGEL, P., NEA Data Bank, Bâtiment 45, F-91191 Gif-sur-Yvette, France

DIAZ MUNOZ, P., NEA Data Bank, Bâtiment 45, F-91191 Gif-sur-Yvette, France

OECD NUCLEAR ENERGY AGENCY
AGENCE DE L'OCDE POUR L'ENERGIE NUCLEAIRE

STADIE, K., Deputy Director, Safety and Regulation, OECD/Nuclear Energy Agency, 38 boulevard Suchet, F-75016 Paris

OLIVIER, J.P., Head, Division of Radiation Protection and Waste Management, 38, boulevard Suchet, F-75016 Paris

NEA SECRETARIAT
SECRETARIAT DE L'AEN

CARLYLE, S.G., Division of Radiation Protection and Waste Management, OECD/Nuclear Energy Agency, 38 boulevard Suchet, F-75016 Paris, France

BRAGG, K., Division of Radiation Protection and Waste Management, OECD/Nuclear Energy Agency, 38 boulevard Suchet, F-75016 Paris, France

SOME OTHER NEA PUBLICATIONS

QUELQUES AUTRES PUBLICATIONS DE L'AEN

Geological Disposal of Radioactive Waste – *An Overview of the Current Status of Understanding and Development* (1984)

Évacuation des déchets radioactifs en formations géologiques – *Panorama de l'état actuel des connaissances* (1984)

£10.00 US$20.00 F100.00

Long-term Radiation Protection Objectives for Radioactive Waste Disposal (1984)

Objectifs de protection radiologique à long terme applicables à l'évacuation des déchets radioactifs (1984)

£9.50 US$19.00 F95,00

Long-term Radiological Aspects of Management of Wastes from Uranium Mining and Milling
(Report by an NEA Group of Experts, 1984)

Aspects radiologiques à long terme de la gestion des déchets résultant de l'extraction et du traitement de l'uranium
(Rapport établi par un Groupe d'experts de l'AEN, 1984)

£12.00 US$24.00 F120,00

Long-term Management of Radioactive Waste – Legal, Administrative and Financial Aspects (1984)

Gestion à long terme des déchets radioactifs – Aspects juridiques, administratifs et financiers (1984)

£7.00 US$14.00 F70,00

Technical Appraisal of the Current Situation in the Field of Radioactive Waste Management (1985)

Appréciation technique sur la situation actuelle en matière de gestion des déchets radioactifs (1985)

Free on request – Gratuit sur demande

Radioactive Waste Disposal – *In Situ Experiments in Granite*
(Proceedings of the 2nd NEA/Stripa Project Symposium, Stockholm 1985)

Évacuation des déchets radioactifs – *Expériences in situ dans du granite*
(Compte rendu du 2e Symposium AEN/Projet de Stripa, Stockholm 1985)

£16.80 US$34.00 F168,00

Storage with Surveillance Versus Immediate Decommissioning for Nuclear Reactors
(Proceedings of the Paris Workshop, 1984)

Stockage sous surveillance ou déclassement immédiat des réacteurs nucléaires
(Compte rendu d'une réunion de travail de Paris, 1984)

£14.00 US$28.00 F140,00

The Management of High-Level Radioactive Waste – *A Survey of Demonstration Activities* (1985)

Gestion des déchets de haute activité – *Bilan des travaux de démonstration* (1985)

Free on request – Gratuit sur demande

OECD SALES AGENTS
DÉPOSITAIRES DES PUBLICATIONS DE L'OCDE

ARGENTINA - ARGENTINE
Carlos Hirsch S.R.L.,
Florida 165, 4° Piso,
(Galeria Guemes) 1333 Buenos Aires
Tel. 33.1787.2391 y 30.7122

AUSTRALIA-AUSTRALIE
D.A. Book (Aust.) Pty. Ltd.
11-13 Station Street (P.O. Box 163)
Mitcham, Vic. 3132 Tel. (03) 873 4411

AUSTRIA - AUTRICHE
OECD Publications and Information Centre,
4 Simrockstrasse,
5300 Bonn (Germany) Tel. (0228) 21.60.45
Local Agent:
Gerold & Co., Graben 31, Wien 1 Tel. 52.22.35

BELGIUM - BELGIQUE
Jean de Lannoy, Service Publications OCDE,
avenue du Roi 202
B-1060 Bruxelles Tel. 02/538.51.69

CANADA
Renouf Publishing Company Limited/
Éditions Renouf Limitée Head Office/
Siège social – Store/Magasin :
61, rue Sparks Street,
Ottawa, Ontario KIP 5A6
Tel. (613)238-8985. 1-800-267-4164
Store/Magasin : 211, rue Yonge Street,
Toronto, Ontario M5B 1M4.
Tel. (416)363-3171
Regional Sales Office/
Bureau des Ventes régional :
7575 Trans-Canada Hwy., Suite 305,
Saint-Laurent, Quebec H4T 1V6
Tel. (514)335-9274

DENMARK - DANEMARK
Munksgaard Export and Subscription Service
35, Nørre Søgade, DK-1370 København K
Tel. +45.1.12.85.70

FINLAND - FINLANDE
Akateeminen Kirjakauppa,
Keskuskatu 1, 00100 Helsinki 10 Tel. 0.12141

FRANCE
OCDE/OECD
Mail Orders/Commandes par correspondance :
2, rue André-Pascal,
75775 Paris Cedex 16
Tel. (1) 45.24.82.00
Bookshop/Librairie : 33, rue Octave-Feuillet
75016 Paris
Tel. (1) 45.24.81.67 or/ou (1) 45.24.81.81
Principal correspondant :
Librairie de l'Université,
13602 Aix-en-Provence Tel. 42.26.18.08

GERMANY - ALLEMAGNE
OECD Publications and Information Centre,
4 Simrockstrasse,
5300 Bonn Tel. (0228) 21.60.45

GREECE - GRÈCE
Librairie Kauffmann,
28 rue du Stade, Athens 132 Tel. 322.21.60

HONG KONG
Government Information Services,
Publications (Sales) Office,
Beaconsfield House, 4/F.,
Queen's Road Central

ICELAND - ISLANDE
Snæbjörn Jónsson & Co., h.f.,
Hafnarstræti 4 & 9,
P.O.B. 1131 – Reykjavik
Tel. 13133/14281/11936

INDIA - INDE
Oxford Book and Stationery Co.,
Scindia House, New Delhi I Tel. 45896
17 Park St., Calcutta 700016 Tel. 240832

INDONESIA - INDONESIE
Pdin Lipi, P.O. Box 3065/JKT.Jakarta
Tel. 583467

IRELAND - IRLANDE
TDC Publishers – Library Suppliers
12 North Frederick Street, Dublin 1
Tel. 744835-749677

ITALY - ITALIE
Libreria Commissionaria Sansoni,
Via Lamarmora 45, 50121 Firenze
Tel. 579751/584468
Via Bartolini 29, 20155 Milano Tel. 365083
Sub-depositari :
Ugo Tassi, Via A. Farnese 28,
00192 Roma Tel. 310590
Editrice e Libreria Herder,
Piazza Montecitorio 120, 00186 Roma
Tel. 6794628
Agenzia Libraria Pegaso,
Via de Romita 5, 70121 Bari
Tel. 540.105/540.195
Agenzia Libraria Pegaso, Via S.Anna dei
Lombardi 16, 80134 Napoli. Tel. 314180
Libreria Hœpli,
Via Hœpli 5, 20121 Milano Tel. 865446
Libreria Scientifica
Dott. Lucio de Biasio "Aeiou"
Via Meravigli 16, 20123 Milano Tel. 807679
Libreria Zanichelli, Piazza Galvani 1/A,
40124 Bologna Tel. 237389
Libreria Lattes,
Via Garibaldi 3, 10122 Torino Tel. 519274
La diffusione delle edizioni OCSE è inoltre assicurata dalle migliori librerie nelle città più importanti.

JAPAN - JAPON
OECD Publications and Information Centre,
Landic Akasaka Bldg., 2-3-4 Akasaka,
Minato-ku, Tokyo 107 Tel. 586.2016

KOREA - CORÉE
Pan Korea Book Corporation
P.O.Box No. 101 Kwangwhamun, Seoul
Tel. 72.7369

LEBANON - LIBAN
Documenta Scientifica/Redico,
Edison Building, Bliss St.,
P.O.B. 5641, Beirut Tel. 354429-344425

MALAYSIA - MALAISIE
University of Malaya Co-operative Bookshop
Ltd.,
P.O.Box 1127, Jalan Pantai Baru,
Kuala Lumpur Tel. 577701/577072

NETHERLANDS - PAYS-BAS
Staatsuitgeverij Verzendboekhandel
Chr. Plantijnstraat, 1 Postbus 20014
2500 EA S-Gravenhage Tel. 070-789911
Voor bestellingen: Tel. 070-789208

NEW ZEALAND - NOUVELLE-ZÉLANDE
Government Printing Office Bookshops:
Auckland: Retail Bookshop, 25 Rutland Street,
Mail Orders, 85 Beach Road
Private Bag C.P.O.
Hamilton: Retail: Ward Street,
Mail Orders, P.O. Box 857
Wellington: Retail, Mulgrave Street, (Head Office)
Cubacade World Trade Centre,
Mail Orders, Private Bag
Christchurch: Retail, 159 Hereford Street,
Mail Orders, Private Bag
Dunedin: Retail, Princes Street,
Mail Orders, P.O. Box 1104

NORWAY - NORVÈGE
Tanum-Karl Johan a.s
P.O. Box 1177 Sentrum, 0107 Oslo 1
Tel. (02) 801260

PAKISTAN
Mirza Book Agency
65 Shahrah Quaid-E-Azam, Lahore 3 Tel. 66839

PORTUGAL
Livraria Portugal,
Rua do Carmo 70-74, 1117 Lisboa Codex.
Tel. 360582/3

SINGAPORE - SINGAPOUR
Information Publications Pte Ltd
Pei-Fu Industrial Building,
24 New Industrial Road No. 02-06
Singapore 1953 Tel. 2831786, 2831798

SPAIN - ESPAGNE
Mundi-Prensa Libros, S.A.,
Castelló 37, Apartado 1223, Madrid-28001
Tel. 431.33.99
Libreria Bosch, Ronda Universidad 11,
Barcelona 7 Tel. 317.53.08/317.53.58

SWEDEN - SUÈDE
AB CE Fritzes Kungl. Hovbokhandel,
Box 16356, S 103 27 STH,
Regeringsgatan 12,
DS Stockholm Tel. (08) 23.89.00
Subscription Agency/Abonnements:
Wennergren-Williams AB,
Box 30004, S104 25 Stockholm. Tel. 08/54.12.00

SWITZERLAND - SUISSE
OECD Publications and Information Centre,
4 Simrockstrasse,
5300 Bonn (Germany) Tel. (0228) 21.60.45
Local Agent:
Librairie Payot,
6 rue Grenus, 1211 Genève 11
Tel. (022) 31.89.50

TAIWAN - FORMOSE
Good Faith Worldwide Int'l Co., Ltd.
9th floor, No. 118, Sec.2
Chung Hsiao E. Road
Taipei Tel. 391.7396/391.7397

THAILAND - THAILANDE
Suksit Siam Co., Ltd.,
1715 Rama IV Rd.,
Samyam Bangkok 5 Tel. 2511630

TURKEY - TURQUIE
Kültur Yayinlari Is-Türk Ltd. Sti.
Atatürk Bulvari No: 191/Kat. 21
Kavaklidere/Ankara Tel. 17.02.66
Dolmabahce Cad. No: 29
Besiktas/Istanbul Tel. 60.71.88

UNITED KINGDOM - ROYAUME UNI
H.M. Stationery Office,
Postal orders only:
P.O.B. 276, London SW8 5DT
Telephone orders: (01) 622.3316, or
Personal callers:
49 High Holborn, London WC1V 6HB
Branches at: Belfast, Birmingham,
Bristol, Edinburgh, Manchester

UNITED STATES - ÉTATS-UNIS
OECD Publications and Information Centre,
Suite 1207, 1750 Pennsylvania Ave., N.W.,
Washington, D.C. 20006 - 4582
Tel. (202) 724.1857

VENEZUELA
Libreria del Este,
Avda F. Miranda 52, Aptdo. 60337,
Edificio Galipan, Caracas 106
Tel. 32.23.01/33.26.04/31.58.38

YUGOSLAVIA - YOUGOSLAVIE
Jugoslovenska Knjiga, Knez Mihajlova 2,
P.O.B. 36, Beograd Tel. 621.992

Orders and inquiries from countries where Sales Agents have not yet been appointed should be sent to:
OECD, Publications Service, Sales and Distribution Division, 2, rue André-Pascal, 75775 PARIS CEDEX 16.

Les commandes provenant de pays où l'OCDE n'a pas encore désigné de dépositaire peuvent être adressées à :
OCDE, Service des Publications. Division des Ventes et Distribution. 2. rue André-Pascal. 75775 PARIS CEDEX 16.

69482-03-1986

OECD PUBLICATIONS, 2, rue André-Pascal, 75775 PARIS CEDEX 16 - No. 43607 1986
PRINTED IN FRANCE
(66 86 04 3) ISBN 92-64-02831-5